AF572722

BIOCHEMICAL SOCIETY SYMPOSIA

No. 69

GLYCOGENOMICS: THE IMPACT OF GENOMICS AND INFORMATICS ON GLYCOBIOLOGY

BIOCHEMICAL SOCIETY SYMPOSIUM No. 69
held at the University of York, December 2001

Glycogenomics: the Impact of Genomics and Informatics on Glycobiology

ORGANIZED AND EDITED BY
K. DRICKAMER AND A. DELL

PORTLAND PRESS

Published by Portland Press,
59 Portland Place, London W1B 1QW, U.K.
on behalf of The Biochemical Society
Tel: (+44) 20 7580 5530; e-mail: editorial@portlandpress.com
http://www.portlandpress.com

ISBN 1 85578 154 9 ISSN 0067-8964

British Library Cataloguing in Publication Data
A catalogue record for this book is available from the British Library

Typeset by Portland Press Ltd
Printed in Great Britain by Black Bear Press Ltd, Cambridge

Contents

Preface vii

Abbreviations ix

1 Functional post-translational proteomics approach to study the role of N-glycans in the development of *Caenorhabditis elegans*
By H. Schachter, S. Chen, W. Zhang, A.M. Spence, S. Zhu, J.W. Callahan, D.J. Mahuran, X. Fan, R.D. Bagshaw, Y.-M. She, J. Cesar Rosa and V.N. Reinhold 1

2 Comparative aspects of glycosyltransferases
By C. Breton, H. Heissigerová, C. Jeanneau, J. Moravcová and A. Imberty 23

3 Glycosyltransferases and glycan structures contributing to the adhesive activities of L-, E- and P-selectin counter-receptors
By J.B. Lowe 33

4 New insights into heparan sulphate biosynthesis from the study of mutant mice
By C.L.R. Merry and J.T. Gallagher 47

5 Genomic analysis of C-type lectins
By K. Drickamer and A.J. Fadden 59

6 Lectins and protein traffic early in the secretory pathway
By H.-P. Hauri, O. Nufer, L. Breuza, H. Ben Tekaya and L. Liang 73

7 New I-type lectins of the CD33-related siglec subgroup identified through genomics
By P.R. Crocker and J. Zhang 83

8 Roles of galectins *in vivo*
By F. Poirier 95

9 MS screening strategies: investigating the glycomes of knockout and myodystrophic mice and leukodystrophic human brains
By M. Sutton-Smith, H.R. Morris, P.K. Grewal, J.E. Hewitt, R.E. Bittner, E. Goldin, R. Schiffmann and A. Dell 105

10 The glycomes of *Caenorhabditis elegans* and other model organisms
By S.M. Haslam, D. Gems, H.R. Morris and A. Dell 117

11 Custom microarray for glycobiologists: considerations for glycosyltransferase gene expression profiling
By E.M. Comelli, M. Amado, S.R. Head and J.C. Paulson 135

12 Neoglycoconjugates: trade and art
By N.V. Bovin 143

Subject index 161

Preface

The sugars that decorate the surfaces of proteins and cells represent a unique form of information used in communication within and between cells. The field of glycobiology addresses the roles of these glycans in protein trafficking, cell–cell adhesion, organization of the extracellular matrix, signalling and other processes. To date, these roles have been studied largely on a case-by-case basis, but the availability of complete genome sequences provides an opportunity to take a global view of the field. The goal of the Biochemical Society's Annual Symposium, which was held at the Society's meeting at the University of York in December 2001, was to assess the special challenges of connecting genomic information with the biochemical processes of glycobiology.

There are three broad aspects to glycobiology: the nature of glycans, the pathway for synthesis of the glycans and the effects the glycans have, either directly on protein function or through lectin-type receptors that bind to them. Because sugars are not encoded directly in genomes, the enzymes that assemble complex oligosaccharides and the lectins that recognize them provide essential links between genomes and glycobiology. The first four chapters of this book focus on what we can learn about the biological functions of glycans by studying biosynthetic enzymes. The subsequent four chapters discuss what we can learn from study of the lectins, while the final four chapters address the special tactics that are needed to develop an overview of what glycans are present in an organism and what their functions are.

In each of these areas, speakers at the symposium were urged to consider three general questions. What is the current state of our understanding and how much more remains to be understood? What can we learn from large-scale analysis based on genomics, high-throughput glycan analysis, screening of sugar-binding activity with glycan arrays and other approaches to a global view of glycobiology? What other ways forward do we envision in the field? Thus, the chapters in this volume illustrate what can be learned from these approaches, but also provide speculation on how the field of glycobiology will develop in the future.

It emerged from the symposium that one important function of genomics is to provide an organizing framework for thinking about glycobiology; for example, the key enzymes in glycan biosynthesis are the glycosyltransferases. The presentations indicated that we have most of the information needed to identify all of the glycosyltransferases in an organism once its genomic sequence is known. Furthermore, tools will soon be available to establish where and when these enzymes are expressed. To complement this, rapid and sensitive analytical methods employing MS will establish what sugars are found where. Thus, it can be envisioned that tissue-by-tissue catalogues of glycosyltransferase expression and glycan structures can be established. It will be interesting to see if correlation of these two large data sets can be achieved.

The same approaches will also be applicable to the lectins that recognize glycans and mediate many of their biological functions. The genome sequences are revealing many candidate lectins that have not been identified previously. Analysis of these sequences suggests possible ligands, and novel expression and screening methods provide the means to identify endogenous high-affinity ligands. The great diversity of lectin structures means that sequence analysis cannot be used to identify all classes of sugar-binding proteins, but those that can be identified will provide important leads for further study.

Genomics is thus a source of ideas about what glycans might be present in an organism and what their function may be. For example, the discovery of a novel glycosyltransferase suggests that there may be a novel biosynthetic pathway; the gene can then provide a route to studying the expression of the enzyme and characterization of its enzymic activity. Perhaps even more importantly, when a glycosyltransferase or lectin gene can be linked to a disease condition, hypotheses about the role of specific glycans can be developed and then tested. Genomics promises to increase the rate at which such links are demonstrated.

There was general agreement that genomics will be a very useful tool in glycobiology; high-sensitivity glycan analysis will speed up discovery and databases will provide valuable catalogues of what is known. However, an important point to emerge from the presentations and discussions at the symposium is that these approaches will not in themselves address the central issue of connecting glycan structure to function. The classical, case-by-case approach is still needed to demonstrate these relationships. This conclusion is particularly evident in studies with knockout mice. These studies have the power to illuminate the biological roles of glycans, but the effects of knocking out a glycosyltransferase or lectin gene can be subtle and complex.

The symposium also highlighted both the advantages and limitations of working with simpler model organisms. The ability to identify glycosyltransferase orthologues in mammals and invertebrates such as *Drosophila melanogaster* and *Caenorhabditis elegans* provides evidence for what can be learned about common functions of glycans throughout the animal kingdom. Some of the intracellular trafficking functions are conserved even in unicellular eukaryotes such as yeasts; however, studies of glycans in invertebrates and analysis of the potential lectin genes in these species suggest a major divergence in extracellular glycans and lectins between invertebrates and vertebrates. In some cases, this divergence will limit the value of invertebrates as models for mammalian biology, but the differences also provide insights into the unique biology associated with various groups of organisms.

We would like to thank the staff of the Biochemical Society for their efforts in organizing the symposium in York and in the preparation of this book. In addition, it is a pleasure to acknowledge the generous financial support of GlycoMinds Ltd, Nestlé, Proteome Systems Ltd, Glyko Inc, Dionex and Omicron Biochemicals.

Kurt Drickamer and *Anne Dell*

Abbreviations

BGT	T4 β-glucosyltransferase
CACH	childhood ataxia with central nervous system hypomyelination
CAD	collision-activated dissociation
CAZY	carbohydrate-active enzyme
CHO	Chinese hamster ovary
COP	coat protein
CRD	carbohydrate-recognition domain
CTLD	C-type lectin-like domain
DG	dystroglycan
2D–IEF–SDS/PAGE	two-dimensional isoelectric focusing–SDS/PAGE
ER	endoplasmic reticulum
ERGIC	ER–Golgi intermediate compartment
ERGL	ERGIC-53-like
ESI–IT–MS	electrospray ionization–ion trap–MS
ESI–Q–TOF–MS	electrospray ionization–quadrupole–time-of-flight–MS
EST	expressed sequence tag
FAB	fast atom bombardment
Fuc	fucose
FucT-VII	α1-3fucosyltransferase VII
Gal	galactose
GalNAc	*N*-acetylgalactosamine
GlcA	glucuronic acid
GlcN	glucosamine
GlcNAc	*N*-acetylglucosamine
GlcNAcT	GlcNAc transferase
GlcNAcT I	UDP-N-acetylglucosamine:α3-D-mannoside β1,2-N-acetylglucosaminyltransferase I
GlcNAcT I.2	UDP-N-acetylglucosamine:α-D-mannoside β1,2-N-acetylglucosaminyltransferase I.2
GlcNAcT II	UDP-N-acetylglucosamine:α6-D-mannoside β1,2-N-acetylglucosaminyltransferase II
GlcNAcT V	UDP-N-acetylglucosamine:α6-D-mannoside β1,6-N-acetylglucosaminyltransferase V
GlcNS	N-sulphylglucosamine
GlyCAM	glycosylation-dependent cell adhesion molecule
GnT	N-acetylglucosaminyltransferase

GST	galactose/N-acetylgalactosamine/N-acetylglucosamine 6-O-sulphotransferase
GT	glycosyltransferase
HEV	high-endothelial venule
Hex	hexose
HexA	hexuronic acid
HexNAc	*N*-acetylhexosamine
HS	heparan sulphate
HS-pol	heparan sulphate polymerase
hSiglec	human Siglec
HSPG	heparan sulphate proteoglycan
IdoA	iduronic acid
ITAM	immunoreceptor tyrosine-based activation motif
ITIM	immunoreceptor tyrosine-based inhibitory motif
ITSM	immunoreceptor tyrosine-based switch motif
LSST	L-selectin ligand sulphotransferase
MAG	myelin-associated glycoprotein
Man	mannose
MEB disease	muscle–eye–brain disease
mSiglec	murine Siglec
MS/MS	tandem MS
NeuAc	*N*-acetylneuraminic acid
NeuGc	*N*-glycolylneuraminic acid
NDST	N-deacetylase/N-sulphotransferase
NK cell	natural killer cell
OST	O-sulphotransferase
PAA	polyacrylamide
PC	phosphorylcholine
PE	phosphatidylethanolamine
PNGase	peptide N-glycosidase
PSGL	P-selectin glycoprotein ligand
Q–TOF	quadrupole–time-of-flight
SHP	Src homology 2-domain-containing protein tyrosine phosphatase
SI	sucrase isomaltase
SLAM	signalling lymphocyte activation molecule
UTR	untranslated region
VIP	vesicular integral membrane protein
WormPD	Worm Proteome Database

Biochem. Soc. Symp. **69**, 1–21
(Printed in Great Britain)

1

Functional post-translational proteomics approach to study the role of N-glycans in the development of *Caenorhabditis elegans*

Harry Schachter*†[1], Shihao Chen*‡, Wenli Zhang*, Andrew M. Spence§, Shaoxian Zhu§, John W. Callahan*†¶, Don J. Mahuran*|, Xiaolian Fan*, Rick D. Bagshaw*, Yi-Min She*, J. Cesar Rosa††‡‡ and Vernon N. Reinhold††**

*Department of Structural Biology and Biochemistry, The Hospital for Sick Children, 555 University Avenue, Toronto, ON, Canada M5G 1X8, †Department of Biochemistry, University of Toronto, 1 King's College Circle, Toronto, ON, Canada M5S 1A8, ‡Burnham Institute, 10901 North Torrey Pines Road, La Jolla, CA 92037, U.S.A., §Department of Molecular and Medical Genetics, University of Toronto, 1 King's College Circle, Toronto, ON, Canada M5S 1A8, ¶Department of Paediatrics, University of Toronto, 1 King's College Circle, Toronto, ON, Canada M5S 1A8, |Department of Medicine, University of Toronto, 1 King's College Circle, Toronto, ON, Canada M5S 1A8, **Department of Laboratory Medicine and Pathobiology, University of Toronto, 1 King's College Circle, Toronto, ON, Canada M5S 1A8, ††Department of Chemistry, University of New Hampshire, Durham, NH 03824, U.S.A. and ‡‡Protein Chemistry Center, Faculty of Medicine of Ribeirao Preto, University of Sao Paulo, Av. Bandeirantes, 3900, 14049-900 Ribeirao Preto/SP, Brazil

Abstract

Glycosylation is one of the most common post-translational protein modifications. Carbohydrate-mediated interactions between cells and their environment are important in differentiation, embryogenesis, inflammation, cancer and metastasis and other processes. Humans and mice with mutations

[1]To whom correspondence should be addressed at Department of Structural Biology and Biochemistry, The Hospital for Sick Children, 555 University Avenue, Toronto, ON, Canada M5G 1X8 (e-mail harry@sickkids.on.ca).

that prevent normal N-glycosylation show multi-systemic defects in embryogenesis, thereby proving that these molecules are essential for normal development; however, a large number of proteins undergo defective glycosylation in these human and mouse mutants, and it is therefore difficult to determine the precise molecular roles of specific N-glycans on individual proteins. We describe here a 'functional post-translational proteomics' approach that is designed to determine the role of N-glycans on individual glycoproteins in the development of *Caenorhabditis elegans*.

Developmental glycobiology: protein-bound glycans are essential for normal metazoan development

The importance of protein-bound glycans in the development of metazoan animals is now widely recognized. Mutations in the genes encoding lysosomal glycosidases [1,2] and UDP-*N*-acetylglucosamine (GlcNAc): lysosomal-enzyme GlcNAc-1-phosphotransferase (which is required for targeting of many glycosidases to the lysosomes) [3–5] have long been known to cause various congenital lysosomal storage diseases. Other examples of congenital diseases with defective protein glycosylation are the progeroid form of Ehlers–Danlos syndrome [6,7], leucocyte adhesion deficiency type II [8] and the congenital disorders of glycosylation [9–12]. Six type I and four type II congenital disorders of glycosylation have been described to date.

Phenotypes have been determined for mice with targeted or natural mutations in the genes that encode UDP-GlcNAc:α3-D-mannoside β1,2-N-acetylglucosaminyltransferase I (GlcNAcT I) [13,14], UDP-GlcNAc:α6-D-mannoside β1,2-N-acetylglucosaminyltransferase II (GlcNAcT II) [15,16], UDP-GlcNAc:β4-D-mannoside β1,4-N-acetylglucosaminyltransferase III [17–19], UDP-GlcNAc:α6-D-mannoside β1,6-N-acetylglucosaminyltransferase V (GlcNAcT V) [20,21], UDP-Gal:GlcNAc β1,4-galactosyltransferase I [22–24], α3,6-mannosidase II [25,26], α1,3-fucosyltransferase VII [27–29], O-GlcNAc-transferase [30], the protein encoded by the Large gene [31] and other glycosyltransferases [32–39].

In humans and mice, the phenotypes of these various mutations range from embryonic lethality, to severe multi-systemic developmental abnormalities, to relatively mild defects in the immune and other systems. The GlcNAcT I null mutation has relatively little effect on viability or growth in Chinese hamster ovary (CHO) and baby hamster kidney cells [40–43], but is lethal to the mouse embryo [13,14], which indicates that hybrid and complex N-glycans are essential for cell–cell interactions and development in multi-cellular animals.

Defective protein glycosylation has also been shown to result in abnormal development in *Drosophila melanogaster* [44,45] and *Caenorhabditis elegans*. The *sqv* (squashed vulva) genes are a set of eight independent loci in *C. elegans*, which are required for the invagination of vulval epithelial cells, normal oocyte formation and embryogenesis [46,47]. The *sqv-3*, *sqv-7* and *sqv-8* genes all affect the biosynthesis of glycosaminoglycans [48]. The *sqv-7* gene encodes a multi-transmembrane hydrophobic protein that can transport several nucleotide sugars [UDP-glucuronic acid, UDP-*N*-acetylgalactosamine,

UDP-galactose (UDP-Gal)] into the Golgi lumen for the biosynthesis of glycoconjugates that are required for development [49]. The *sqv-3* gene is homologous with the human gene that encodes UDP-Gal:xylose-β1-O-Ser-R β1,4 galactosyltransferase I, which is involved in the synthesis of the glycosaminoglycan–protein linkage region of proteoglycans [50]; a defect in this enzyme has been shown to be the cause of the progeroid form of Ehlers–Danlos syndrome [6,7]. The *C. elegans* genome contains at least 100 genes that are homologous with genes that are mutated in human disease [51], which suggests that other human disease genes causing abnormal glycosylation may have *C. elegans* homologues.

It is clear therefore that interference with normal glycosylation disrupts development in humans, mouse and invertebrates, indicating that protein-bound glycans play essential roles in metazoan development.

Function of specific glycans on specific glycoproteins

In order to understand exactly what a protein-bound glycan is doing at the molecular level, it is essential to understand its precise role in the function of a specific protein. There are several excellent examples that provide this type of information in the literature.

The congenital lysosomal storage disease mucolipidosis II (inclusion cell disease or I-cell disease) and a milder form named mucolipidosis III (Pseudo–Hurler polydystrophy) provide such an example [1]. These diseases are caused by multiple deficiencies in lysosomal enzymes, which result due to defective targeting of these enzymes to the lysosome [5]. The defective gene encodes UDP-GlcNAc:lysosomal-enzyme GlcNAc-1-phosphotransferase which transfers GlcNAcα1-O-phosphate to C-6 of the mannose residues of certain oligomannose N-glycans on lysosomal hydrolases. The GlcNAc is subsequently removed by GlcNAc-1-phosphodiester α-N-acetylglucosaminidase [3,52] to expose the mannose 6-phosphate signal required for recognition by the mannose 6-phosphate receptors and subsequent lysosome targeting. The phospho-GlcNAc-transferase has been cloned [4,53,54] and the defects responsible for improper targeting in mucolipidosis III have been determined [4].

The E-, P- and L-selectins are proteins with lectin domains that bind to carbohydrate moieties on their respective glycoprotein ligands. The interactions between the selectins and their ligands play important roles in various aspects of the immune system such as lymphocyte homing and the extravasation of lymphocytes from the capillary lumen into the tissues at sites of inflammation. Although details remain to be resolved, the mechanisms by which the selectins function are now reasonably well understood [28,55–58].

Two recent examples of great interest to developmental glycobiologists involve oligosaccharides that are linked to protein by Fucα1-*O*-Ser/Thr or Manα1-*O*-Ser/Thr. Notch is a large cell-surface receptor, which is known to be an essential player in a wide variety of developmental cascades. Fringe proteins can positively and negatively modulate the ability of Notch ligands to activate the Notch receptor. The *Drosophila* and mammalian genes that encode Fringe proteins have been cloned and the proteins have been shown to be UDP-

GlcNAc:Fucα1-O-Ser/Thr-R β1,3-N-acetylglucosaminyltransferases that initiate elongation of O-linked fucose residues attached to epidermal growth factor-like sequence repeats of Notch [59,60]. Fringe enzymic activity is essential for its modulatory action. The enzyme that adds *O*-fucose to epidermal growth factor-like repeats, GDP-Fuc:protein O-fucosyltransferase, has also been cloned [61]. The O-linked oligosaccharide in CHO cells has the structure Sialylα2,3Galβ1,4GlcNAcβ1,3Fucα1-*O*-Ser/Thr [62]. The trisaccharide Galβ1,4GlcNAcβ1,3Fucα1-*O*-Ser/Thr is the minimal O-fucosylated glycan which can support Fringe modulation of Notch signalling [63].

Dystroglycan (dystrophin-associated glycoprotein) is a cytoskeleton-linked extracellular matrix receptor that is expressed in many cell types and is composed of α- and β-subunits that are encoded by a single mRNA [64]. Dystroglycan is a central component of the dystrophin–glycoprotein complex, a protein assembly that plays a critical role in a variety of muscular dystrophies [65]. The major sialylated O-glycosidically linked oligosaccharide of bovine peripheral nerve [66] and rabbit skeletal muscle [67] α-dystroglycan is a novel O-mannosyl-type oligosaccharide, Sialylα2,3Galβ1,4GlcNAcβ1,2Manα1-*O*-Ser/Thr [68]. This oligosaccharide is probably required for the binding of laminin to α-dystroglycan. A human enzyme homologous with GlcNAcT I, UDP-GlcNAc:α-D-mannoside β-1,2-N-acetylglucosaminyltransferase I.2 (GlcNAcT I.2) [69–73], appears to be responsible for the synthesis of the GlcNAcβ1,2Manα1-*O*-Ser/Thr moiety of α-dystroglycan since six patients with muscle-eye-brain disease [MIM 253280 (Mendelian Inheritance in Man database; www.ncbi.nlm.nih.gov)] have been shown recently to have mutations in the GlcNAcT I.2 gene [73]. Muscle–eye–brain disease is an autosomal recessive disease of previously unknown aetiology, which is characterized by severe mental retardation, ocular abnormalities, congenital muscular dystrophy and a polymicrogyria-pachygyria-type neuronal migration disorder of the brain.

Challenges of post-translational modifications

The above examples of glycan function are, of course, most interesting and exciting; however, in 1999, Apweiler et al. [74] estimated that 65–69% of the proteins in all of the metazoan databases contain at least one potential N-glycosylation consensus sequence (sequon Asn-Xaa-Ser/Thr where Xaa can be any amino acid except proline) and thus may be N-glycosylated. The number of proteins filed as N-glycoproteins at the time was 7942, but the authors concluded that at least half of all metazoan proteins probably contain N-glycans. The number of proteins with other types of glycans is no doubt equally large. To make matters even more complicated, there are hundreds of other types of post-translational modifications [75,76]. Will it ever be possible to determine the functions of the thousands of proteins that are present in every metazoan genome by the various ‘functional proteomics’ approaches now being widely used and promoted (e.g. comparison of different tissues or biological states by micro-array technology; determination of protein binding patterns)? Even more problematic, will it ever be possible to determine the role that a specific post-translational modification plays in the function of a specific protein for every protein in the genome? Even if one assumes that the lessons learned for a particular modification on a particular protein can probably be extrapolated to groups of other similar proteins, the task seems impossibly difficult.

The function of a gene and the protein it encodes can be studied in various ways. The most direct method is to study the function of a natural or recombinant form of the protein using *in vitro* assays. The role of post-translational modifications can then be determined. Alternatively, one can use cells in culture in a variety of approaches to study gene and protein function, e.g. Stanley and colleagues [77–79] have made major contributions to our understanding of glycan function by developing a series of lectin-resistant mutant CHO cell lines. There are some limitations to the use of such somatic cell mutants. Cell lines deficient in GlcNAcT I, for example, cannot make hybrid or complex N-glycans, but nevertheless have an essentially normal phenotype [40–43]. Similar results have been obtained with somatic cell mutants that are unable to make various types of proteoglycans [80] and O-glycans [81]. One must therefore use animal models if one wishes to determine the functions of glycans that are not essential for cell viability and growth, or to study the role of glycans in development and morphogenesis.

Mutations in genes that affect glycosylation have been studied in humans, mice and invertebrates (see above) and have provided a large body of information on glycan function. An excellent review by Stanley [82] outlined the many new insights obtained by the study of transgenic mice that mis-express or over-express a particular glycosyltransferase and of mice with targeted mutations in a glycosyltransferase gene. The author concluded that "it is now apparent that every glycosylation-defective mouse has been enormously informative"; however, the author also pointed out that "identifying the molecular basis of an altered glycosylation phenotype is difficult because glycosyltransferases modify many substrates and because related glycans are expressed on different glycoconjugates". A transgenic or 'knockout' mouse model can certainly be used to identify some of the protein targets which are acted on by the glycosyltransferase being studied. The role of the glycan in the function of such a protein can then be studied by *in vitro* assays using proteins in which glycosylation has been altered (e.g. by treatment with glycosidases, protein expression in the presence of glycosylation inhibitors, or expression of proteins with mutations that alter glycosylation).

Proteomics

The proteome defines the protein complement of a cell or tissue in a given state at a given time. Proteomics is the science of determining the nature of the proteome and it can be divided into structural and functional proteomics. There is a tendency to equate proteomics with rapid, high-throughput technology and indeed, such approaches are highly desirable, if possible. Functional proteomics usually means quantitative comparisons of different cells or tissues in different states and/or at different times by various high-throughput, high-resolution separation methods, DNA and protein micro-arrays, and protein-binding assays.

Post-translational modifications create major complications for the field of proteomics (see above). There have been several recent reports on the application of the proteomics approach to this problem. Protein glycosylation and phosphorylation have been studied by selective identification of post-translational modifications with modification-specific fragment ions produced by collision-induced dissociation during analysis of peptides by liquid chromatography–electrospray–MS [83,84]. The nano-electrospray ion source (NanoES) has been used for similar studies [85]. The phosphate moieties of phosphoserine/threonine-containing peptides can be replaced chemically by affinity tags and the tags can be used to enrich the phosphopeptides prior to MS analysis [86,87]. A method has been developed that utilizes phosphoprotein isotope-coded affinity tags, which combines stable isotopes and biotin-labelling to enrich and quantitatively measure differences in the O-phosphorylation states of proteins [88]. Aebersold's group [89] has described a rapid and general method for the analysis of protein phosphorylation in complex protein mix-

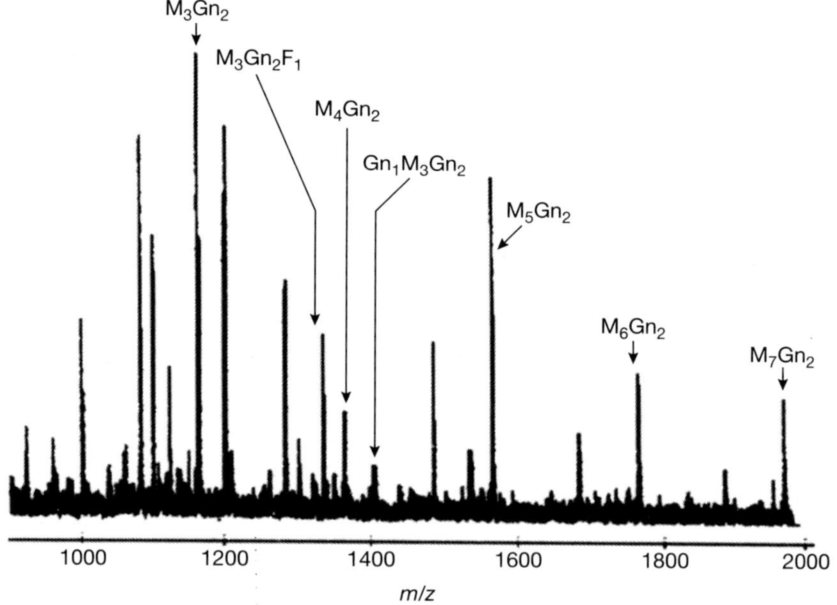

Figure 1 Mass spectrum of N-glycans from *C. elegans*. Protein extracts were obtained from wild-type worms and from worms with null mutations in either *gly-12* or *gly-14* or from *gly-12/gly-14* double null mutants. The genes *gly-12* and *gly-14* encode proteins with GlcNAcT I activity [98–100]. Protein extracts were quantified using Coomassie Blue staining and identical amounts of protein were reduced, alkylated and subjected to trypsin digestion. The peptides were purified by microtips packed with Vydac C_{18} resin. N-glycans were released by protein N-glycanase A in 1 mM sodium acetate (pH 5.0) or by PNGase F in 100 mM ammonium bicarbonate (pH 7.8) at 37°C overnight. The released N-gly-

tures involving the following: (i) selective phosphopeptide isolation from a peptide mixture via a sequence of chemical reactions; (ii) phosphopeptide analysis by automated liquid chromatography–tandem MS (LC-MS/MS); and (iii) identification of the phosphoprotein and the phosphorylated residue(s) by correlation of tandem MS data with sequence databases. Structural analysis of glycans on individual proteins has, of course, been carried out by NMR [90] and MS [91–94] for many years.

These reports indicate that structural post-translational proteomics is a promising area for future studies. The next hurdle is functional post-translational proteomics. We consider this term to mean a general approach to determining the function of a specific post-translational modification on a specific protein. This definition implies a protein-by-protein approach and therefore rules out high-throughput technology. In the remainder of this chapter, we outline a procedure for carrying out functional post-translational proteomics using N-glycans as the post-translational modification under study and *C. elegans* as the model organism.

cans were separated from peptides by reverse-phase C_{18} chromatography, dried by flash evaporation, permethylated and analysed by electrospray–ion trap–MS (ESI–IT–MS). The spectrum from an extract of wild-type worms is shown here. The main components are oligomannose N-glycans ($Hex_{5–7}HexNAc_2$) and paucimannose N-glycans ($Hex_{3–4}HexNAc_2$ and $Hex_3HexNAc_2Fuc_1$). The paucimannose N-glycans are 'footprints' of GlcNAcT I action since they usually require the actions of the GlcNAcT I-dependent enzymes α3,6-mannosidase II and core Fuc-transferase (see text and legend to Figure 2). A small peak of truncated hybrid N-glycan ($Hex_3HexNAc_3$, which is presumed to be Manα1,6[GlcNAcβ1,2Manα1,3]Manβ1,4GlcNAcβ1,4GlcNAc; Figure 2) is also observed; this glycan is indicative of GlcNAcT I action. As an analogue to ICAT (isotope-coded affinity tag) technology for protein quantification [124–126], we have evaluated glycan quantification with deuterated analogues [107]. In a method that is comparable with ICAT strategies to circumvent ion suppression and pre-fractionation losses with protein samples, we have utilized methylation of N-glycans with deuterated and non-deuterated iodomethane to provide a relative carbohydrate quantification with glycoprotein mixtures. ESI–IT–MS responses for standard mixtures of deuterated and non-deuterated samples demonstrated an error of less than 5.0%. Eight N-glycan species were identified by their *m/z* values: $Hex_9HexNAc_2$, 2503/2396; $Hex_8HexNAc_2$, 2289/2193; $Hex_7HexNAc_2$, 2076/1989; $Hex_6HexNAc_2$, 1862/1783; $Hex_5HexNAc_2$, 1650/1580; $Hex_3HexNAc_3$, 1477/1417; $Hex_3HexNAc_2Fuc_1$, 1403/1345; $Hex_3HexNAc_2$, 1223/1171 (deuterated/non-deuterated peaks). The mass spectra for the *gly-12* and *gly-14* null worms and the *gly-14/gly-12* double null mutant worms (not shown) were identical to the wild-type spectrum. The mutant worm spectra were compared with the wild-type worm spectrum using deuterated and non-deuterated samples. For all three mutant worms, the ratios of wild-type (deuterated) to mutant (non-deuterated) samples for all eight glycan peaks identified by MS were almost identical to the ratios of wild-type (deuterated) to wild-type (non-deuterated) control samples (results not shown). M, mannose; Gn, GlaNAc; F, L-fucose.

Is *C. elegans* suitable for the study of N-glycosylation?

C. elegans is considered to be the most completely understood metazoan in terms of anatomy, genetics, development and behaviour [95–97]. The organism is a free-living soil nematode worm, approx. 1 mm in length, with a life cycle of approx. 3 days. The locations, characteristics and lineages of all the somatic cells in the organism and a complete 'wiring diagram' of cell contacts have been determined and the complete genomic DNA sequence is available. The study of mutant alleles of developmentally important genes has provided detailed information on the developmental biology of the worm. The haploid genome contains 10^8 nucleotide pairs with approx. 18000 genes. Worm mutants are readily obtained by chemical or irradiation mutagenesis. Cloned genes can be reintroduced into the *C. elegans* germ line by microinjection of DNA into the syncytial ovary of the hermaphrodite worm or into oocyte nuclei. The injected DNA usually forms extrachromosomal tandem arrays. Reintroduced genes are expressed and can suppress (rescue) a mutant phenotype. Selectable genetic markers are used to identify transformed worms.

The *C. elegans* genomic DNA and Expressed Sequence Tag databases contain sequences that show significant similarities to a large number of enzymes involved in both N- and O-glycan synthesis. *C. elegans* has three GlcNAcT I genes (*gly-12*, *gly-13*, *gly-14*) [98]. We have used UV irradiation in the presence of trimethylpsoralen as a mutagenic agent to obtain worms with null mutations in *gly-12*, *gly-13* and *gly-14* as well as a *gly-14/gly-12* double null mutant [99,100]. The *gly-12*, *gly-14* and *gly-14/gly-12* null mutants all had apparently normal phenotypes [99,100] and a wild-type pattern of N-glycan composition, as revealed by MS (Figure 1). While GLY-12 and GLY-14 can both transfer GlcNAc to Manα1,6(Manα1,3)Manβ1-*O*-octyl(Man3-octyl), GLY-13 acts only on [Manα1,6(Manα1,3)Manα1,6][Manα1,3]Manβ1-*O*-R acceptors (Man_5-R) (Table 1). Assays of extracts of wild-type worms showed a very low GlcNAcT I activity with Man3-octyl and a relatively high activity with Man_5-R (Table 1). Extracts of the *gly-14/gly-12* double null mutant showed no detectable activity with Man3-octyl, but the same activity as wild-type worms when Man_5-R was used as an acceptor substrate (Table 1), leading to the conclusion that GLY-13 is the major GlcNAcT I enzyme in the wild-type worm. Consistent with this conclusion is the finding that most *gly-13* null worms die at the L1 stage of development; however, a small percentage of these worms survive, but they have a starved appearance, frequent morphological abnormalities, egg-laying defects, and reproduce at a much reduced rate. The lethal mutation has been mapped to within approx. 8 kb on either side of the *gly-13* gene, indicating a high probability that lethality is caused by the lack of *gly-13* expression.

Although there is a great deal of variation between different plants, several plant species (almond, hazelnut, pistachio, potato, tomato, walnut) have been shown to accumulate over 30% of their N-glycans as the paucimannose Man_3-R compound Manα1, 6(Manα1,3)(Xylβ1,2) Manβ1,4GlcNAcβ1,4(Fucα1,3)GlcNAc-Asn-Xaa [101]. Since both the β1,2-xylosyltransferase and the core α1,3-fucosyltransferase require prior action of GlcNAcT I [102], the formation of these paucimannose compounds indicates that

Table 1 GlcNAcT I activities in extracts from normal *C. elegans*, transgenic worms over-expressing *gly-12*, *gly-13* and *gly-14* under the control of heat-shock promoter, and worms with null mutations in *gly-12* and *gly-14*. All enzyme assays are averages of duplicate determinations. Every value under either the Man3-octyl or Man_5-R column represents a different worm isolate. *Manα1,6(Manα1,3)Manβ1-O-octyl as acceptor [98]. †[Manα1,6(Manα1,3)Manα1,6][Manα1,3]Manβ1,4GlcNAcβ1,4GlcNAc-Asn-Xaa as acceptor. ‡GLY-14 expressed in the baculovirus/Sf9 system was shown to be active with Man_5-R acceptor [98].

Worms	GlcNAcT I activity ($nmol \cdot mg^{-1} \cdot h^{-1}$) Man3-octyl*	Man_5-R†
Wild-type	0.096, 0.074	4.1, 5.5, 11.1
Heat-shocked over-expressing worms		
GLY-12	2.6, 15.1	Not tested
GLY-13	0.07, 0.09, 0.1, 0.2, 0.4	22, 46, 108, 140
GLY-14	3.7, 4.3, 10.2, 17.5	Active‡
Null mutant worms		
gly-12	0.017	
gly-14	0.063	
gly-14/ *gly-12*	<0.03	8.1, 11.1

a β-N-acetylglucosaminidase, which is presumably located in the Golgi apparatus, removes the GlcNAcβ1,2 residue from the N-glycan core while the nascent glycoprotein traverses the lumen of the endoplasmic reticulum–Golgi assembly line [103,104]. Similarly, insects have been shown to accumulate the paucimannose Manα1,6(Manα1,3)Manβ1,4GlcNAcβ1,4(±Fucα1,3)GlcNAc-Asn-Xaa glycans and to have a specific membrane-bound β-N-acetylglucosaminidase which presumably acts within the lumen of the assembly line during N-glycan biosynthesis [105,106].

We have obtained evidence for a similar mechanism in *C. elegans*. MS analyses [92,107–110] of wild-type *C. elegans* extracts show the presence of oligomannose N-glycans ($Hex_{5-10}HexNAc_2$) and paucimannose N-glycans ($Hex_{3-4}HexNAc_2$ and $Hex_{3-4}HexNAc_2Fuc_1$) (Figure 1). There is also a small MS peak of $Hex_3HexNAc_3$ (Figure 1) that is presumed to be caused by the truncated hybrid structure Manα1,6{GlcNAcβ1,2Manα1,3}Manβ1,4GlcNAcβ1,4GlcNAc (Figure 2) and therefore to be indicative of prior GlcNAcT I action. The accumulation of paucimannose N-glycans in *C. elegans* suggests that worms have a β-N-acetylglucosaminidase similar to the enzyme in plants and insects (Hase; Figure 2). GLY-13 cannot act on the Man_3-R compounds and, therefore, formation of these compounds is expected to be a metabolic 'dead end' in *C. elegans*. Indeed, the peaks for Manα1,6(Manα1,3)Manβ1,4GlcNAcβ1,4(±Fuc)GlcNAc-Asn-Xaa are relatively large (Figure 1). Since previous GlcNAcT I action is essential for the actions of both α3,6-mannosidase II and core FucT (Figure 2), these Man_3-R compounds, either with or without a core Fuc, are believed to be 'footprints' of prior GlcNAcT I action.

In an attempt to provide evidence for this hypothesis, we have carried out MS analyses on N-glycans isolated from a *C. elegans* strain with a null mutation in the *gly-13* gene, which encodes the major GlcNAcT I enzyme in the wild-type worm. N-glycans from wild-type and *gly-13* null worms were

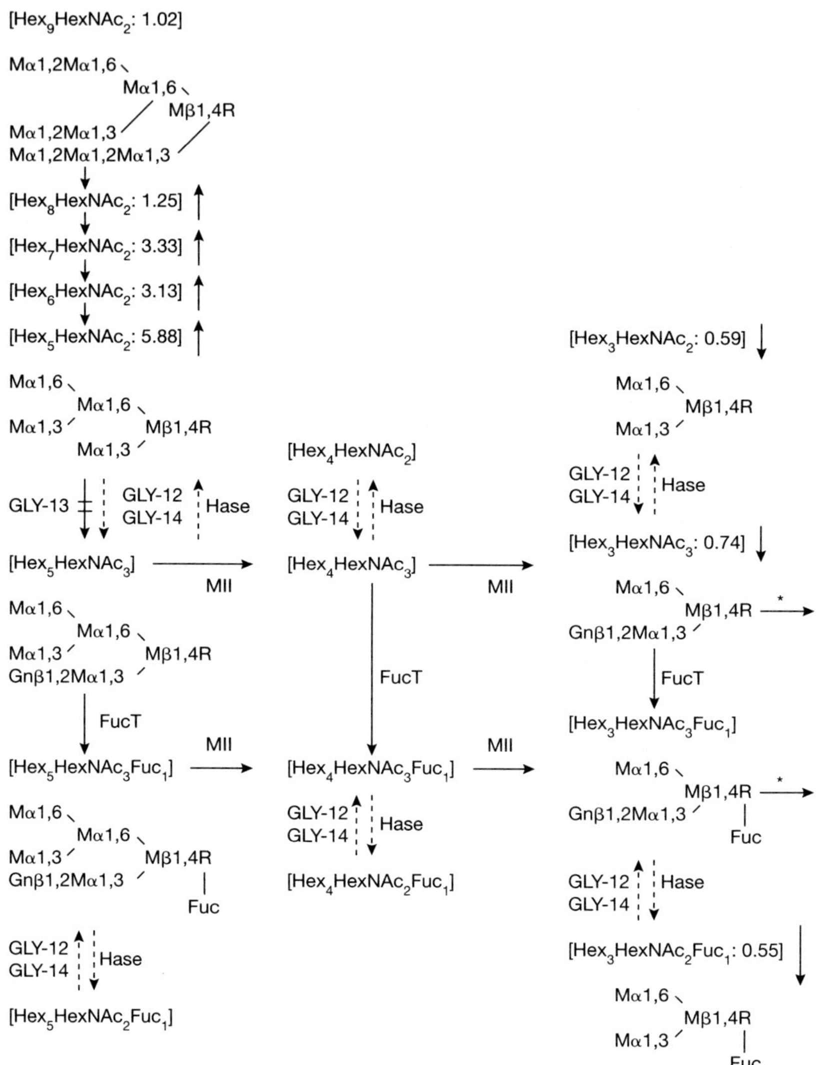

Figure 2 Conversion of oligomannose N-glycans to hybrid and paucimannose N-glycans in wild-type and *gly-13* null mutants of *C. elegans*. This figure shows the conversion of oligomannose N-glycans ($Man_{9\text{–}5}GlcNAc_2$-Asn-Xaa) to hybrid and paucimannose N-glycans (M, mannose; Gn, GlcNAc; Hex, hexose; HexNAc, *N*-acetylhexosamine; R, GlcNAcβ1,4GlcNAc-Asn-Xaa). $Man_9GlcNAc_2$-Asn-Xaa is converted into $Man_5GlcNAc_2$-Asn-Xaa by the sequen-

compared quantitatively by differential methylation with deuterated and non-deuterated iodomethane respectively (see legend to Figure 1). The samples were pooled and analysed by electrospray ionization–ion trap–MS (ESI–IT–MS). Eight N-glycan species were identified by their *m/z* values, and the ratio of the MS peak sizes (*gly-13* null relative to wild-type worms) for each structure was calculated (Figure 2).

tial action of α1,2-mannosidase(s). GlcNAcT I adds a GlcNAc in β1,2 linkage to the Manα1,3- arm of $Man_5GlcNAc_2$-Asn-Xaa to form the hybrid N-glycan $GlcNAc_1Man_5GlcNAc_2$-Asn-Xaa. *C. elegans* has three GlcNAcT I genes *(gly-12, gly-13, gly-14)* [98]. GLY-13 is the major GlcNAcT I enzyme in the wild-type worm (see text). The broken arrows for GLY-12 and GLY-14 indicate that these activities play minor roles in N-glycan synthesis in wild-type worms *in vivo*. Two Man residues are removed from the product of GlcNAcT I by the action of α3,6-mannosidase II (MII) to form the truncated hybrid N-glycan $GlcNAc_1Man_3GlcNAc_2$-Asn-Xaa. Core fucosyltransferase (FucT) may add a Fuc to the Asn-linked GlcNAc, as indicated in the figure. Both α3,6-mannosidase II and FucT require the prior action of GlcNAcT I, and the resulting GlcNAcβ1,2Manα1,3- moiety, for their actions. Wild-type *C. elegans* accumulates the paucimannose N-glycans Manα1,6(Manα1,3)Manβ1,4GlcNAcβ1,4(+/−Fuc) GlcNAc-Asn-Xaa (Figure 1), probably due to the action of a specific β-*N*-acetylglucosaminidase (Hase; broken arrows indicate that the presence of this activity in *C. elegans* has not as yet been proven directly). GLY-13 cannot act on the Man_3-R compounds (see text) and therefore, formation of these compounds is expected to be a metabolic 'dead end' in *C. elegans*. Since prior action of GlcNAcT I is essential for the actions of both α3,6-mannosidase II and core FucT, the paucimannose Man_3-R compounds are 'footprints' of prior GlcNAcT I action. N-glycans from wild-type and *gly-13* null mutant worms were methylated differentially with deuterated and non-deuterated iodomethane and analysed by ESI–IT–MS (see Figure 1). The ratio of the MS peak sizes (*gly-13* null relative to wild-type worms) for each structure is shown after the N-glycan composition. The data indicate that absence of GLY-13 (blocked arrow) results in significant increases in the amounts of $Man_{5–8}GlcNAc_2$-Asn-Xaa structures and significant decreases in Manα1,6(GlcNAcβ1,2Manα1,3) Manβ1,4GlcNAcβ1,4GlcNAc-Asn-Xaa and the paucimannose structure Manα1,6(Manα1,3)Manβ1,4GlcNAcβ1,4(±Fuc)GlcNAc-Asn-Xaa. MS peaks for $Hex_{4–5}HexNAc_3$, $Hex_{4–5}HexNAc_3Fuc_1$ and $Hex_3HexNAc_3Fuc_1$ were not seen ([110]; J. Cesar Rosa and V.N. Reinhold, unpublished work) presumably because rates of removal were significantly more rapid than rates of formation. Although an MS peak for $Hex_5HexNAc_2Fuc_1$ was not detected, Altmann et al. [110] detected an O-methylated version of $Hex_5HexNAc_2Fuc_1$. $Hex_4HexNAc_2$ (Figures 1 and 2) is presumably synthesized after the conversion of $Hex_5HexNAc_3$ into $Hex_4HexNAc_3$ by α3,6-mannosidase II followed by the action of β-N-acetylglucosaminidase. $Hex_4HexNAc_2Fuc_1$ can be formed as indicated in the figure. Several peaks, that are suggestive of N-glycans with more than one Fuc residue, have also been reported in *C. elegans* [110]. Evidence for the synthesis of significant amounts of truncated bi-, tri- and tetra-antennary complex N-glycans by *C. elegans* has been obtained (see text); these compounds are presumably formed by the action of GlcNAcT II at the arrows with the asterisk, followed by the actions of other branching GlcNAc-transferases.

The data in Figure 2 indicate that the absence of GLY-13 results in significant increases in the amounts of oligomannose $Man_{5-8}GlcNAc_2$-Asn-Xaa structures and decreases in the amounts of truncated hybrid $Hex_3HexNAc_3$ [Manα1,6(GlcNAcβ1,2Manα1,3)Manβ1,4GlcNAcβ1,4GlcNAc-Asn-Xaa] and of the paucimannose structures $Hex_3HexNAc_2$ and $Hex_3HexNAc_2Fuc_1$ [Manα1,6(Manα1,3)Manβ1,4GlcNAcβ1,4(±Fuc)GlcNAc-Asn-Xaa]. This pattern proves that synthesis of the paucimannose N-glycans depends, at least in part, on prior GLY-13 action and that *C. elegans*, like plants and insects, must have a β-*N*-acetylglucosaminidase that can act on N-glycans during biosynthesis *in vivo*.

There are, however, detectable levels of $Hex_3HexNAc_3$, $Hex_3HexNAc_2$ and $Hex_3HexNAc_2Fuc_1$ in the *gly-13* null worm (Figure 2). GLY-12 and/or GLY-14 may be at least partly responsible for the synthesis of N-glycans downstream of the GLY-13 block (Figure 2). Interestingly, the *gly-13* null mutation causes a smaller decrease in $Hex_3HexNAc_3$ levels (ratio = 0.74) than in $Hex_3HexNAc_2$ (ratio = 0.59) and $Hex_3HexNAc_2Fuc_1$ (ratio = 0.55) levels (Figure 2). Since there is no activity of GLY-13 on the paucimannose glycans ($Hex_3HexNAc_2$ and $Hex_3HexNAc_2Fuc_1$) in both wild-type and *gly-13* null worms, this shift towards $Hex_3HexNAc_3$ in the *gly-13* null worm suggests a compensatory up-regulation of GLY-12 and/or GLY-14 activity in the absence of GLY-13 and a partial reversal of β-N-acetylglucosaminidase action (Figure 2). The presence of an 'α3,6-mannosidase III-like' enzyme, which can convert $Man_5GlcNAc_2$-Asn-Xaa into $Man_3GlcNAc_2$-Asn-Xaa without prior GlcNAcT I action [25], can explain the formation of $Man_3GlcNAc_2$-Asn-Xaa but not of $Man_3GlcNAc_2Fuc_1$-Asn-Xaa. The effect of lysosomal glycosidases during the worm extraction procedure is difficult to assess; however, extraction was carried out under conditions that minimize lysosomal hydrolase action.

The results described above have important implications. The usefulness of *C. elegans* in studying the role of complex N-glycans in development and other biological processes depends on the worm's ability to make reasonable amounts of these compounds. MS analysis of wild-type worms has shown little or no complex N-glycans and a relatively low level of $GlcNAc_1Man_3GlcNAc_2$-Asn-Xaa, the product of α3,6-mannosidase II action (Figure 1); however, our demonstration that the large MS peaks of Manα1,6(Manα1,3)Manβ1,4GlcNAcβ1,4(±Fuc)GlcNAc-Asn-Xaa are indeed 'footprints' of previous GlcNAcT I action indicate that GlcNAcT I may play essential roles in *C. elegans* development. *C. elegans* has genes encoding enzymically active GlcNAcT II [111], GlcNAcT V [112] and other glycosyltransferases [113,114] required for complex N-glycan synthesis. Recent MS work by Haslam et al. (see Chapter 10 in this volume) has shown the presence in wild-type *C. elegans* of truncated bi-, tri- and tetra-antennary complex N-glycans in which C-6 of the terminal antennary GlcNAc residues is substituted with a phosphorylcholine group. Phosphorylcholine on N-glycans has been described previously [115–118] and it interferes with MS identification of the glycan.

Functional post-translational proteomics approach to study the role of N-glycans in the development of *C. elegans*

This section outlines a general 'functional post-translational proteomics' approach to determine how N-glycans on individual glycoproteins function in the development of *C. elegans*. It is important to point out that this scheme is not meant to replace existing approaches, but rather to complement them. Projects such as the functions of Notch and dystroglycan (see above) will continue to be generated owing to scientific interest in specific glycoproteins. Mutant cell lines, mutant mice and other mutant organisms (*C. elegans*, *Drosophila*, etc.) will continue to provide clues in the study of specific glycoproteins. Our proteomics approach will, hopefully, locate other protein targets for the investigation of glycan function; however, since many proteins may not fulfil the requirements of the method, we have no illusions that our approach is capable of a comprehensive sweep. It should be pointed out that the biological 'read-out' of our screening is a change in the development of a metazoan organism rather than the 'read-outs' that are in common use in functional proteomics at present, i.e. a change in the pattern of gene expression or in the binding of proteins to other proteins.

Step 1: structural post-translational proteomics

We must first identify, on a proteomics scale, those *C. elegans* proteins that carry N-glycans. Worm extracts are subjected to chromatography on lectin columns, followed by two-dimensional isoelectric focusing–SDS/PAGE (2D–IEF–SDS/PAGE) (Figure 3) [119]. Spots are cut from the gel, proteins are extracted and digested with trypsin, and peptides are purified and analysed by either matrix-assisted laser-desorption ionization–time-of-flight-MS or electrospray ionization–quadrupole–time-of-flight–MS (ESI–Q–TOF–MS). The instrument for the latter [120] is capable of peptide sequencing using nano-electrospray tandem MS. We have shown (Figures 1 and 2) that N-glycans carrying the structure GlcNAcβ1,2Manα1,3Manβ1-*O*-R and 'footprints' of GlcNAcT I action can be detected in extracts of *C. elegans*, indicating that GlcNAcT I is functional in the worm and that its products can be detected by MS. ESI–Q–TOF–MS analysis of a tryptic digest of one of the spots shown in Figure 3(B) (circled) is shown in Figure 4(A). ESI–Q–TOF–MS/MS analysis of the $\{M+2H\}^{2+}$ ion at $m/z = 577.31$ (Figure 4A) yielded the peptide sequence Glu-Leu-Gly-Leu-Gly (Figure 4B), which allowed identification of the protein as hypothetical protein T13F3.6 by a search of the *C. elegans* database. MS/MS analysis of the $\{M+2H\}^{2+}$ ion at $m/z = 848.79$ (Figure 4A) indicated the presence of a glycopeptide with a glycan moiety of molecular mass 892.33 Da (Figure 5) representing $Hex_3HexNAc_2$, i.e. the paucimannose structure Manα1,6{Manα1,3}Manβ1,4GlcNAcβ1,4GlcNAc- (Figure 2). Examination of the amino-acid sequence of hypothetical protein T13F3.6 identified two potential N-glycosylation sites (Asn-Xaa-Thr/Ser), one of which (underlined) is in the tryptic peptide Ser-Ile-<u>Asn-Val-Thr</u>-Asp-Arg (theoretical molecular mass = 803.41 Da). Subtracting the molecular mass of the glycan ($\{M+H\}^{1+} = 892.33$ Da) from the total molecular mass ($\{M+H\}^{1+} = 1696.77$ Da)

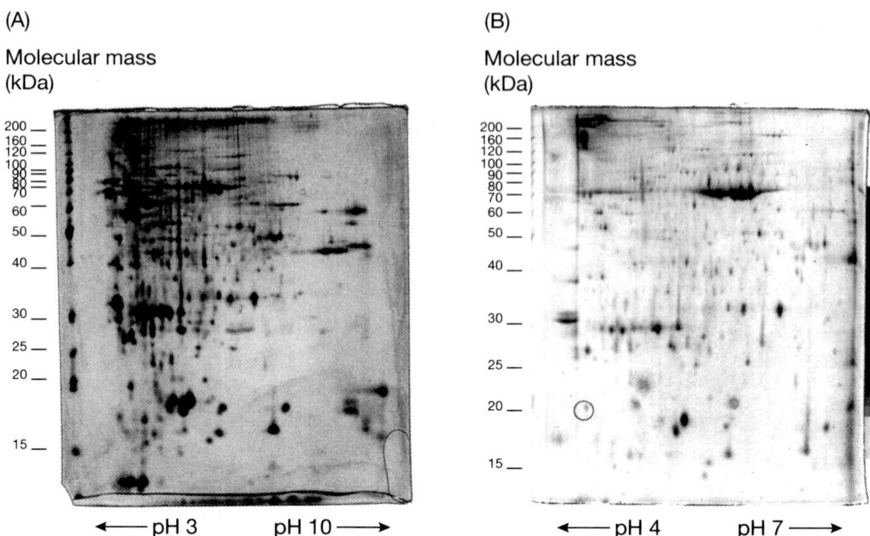

Figure 3 2D–IEF–SDS/PAGE gels of *C. elegans* glycoproteins. Worm extracts were centrifuged to remove membranous organelles and the supernatants were subjected to chromatography on Concanavalin A columns followed by 2D–IEF–SDS/PAGE [119]. (**A**) Glycoproteins (0.4 mg) were first separated on a non-linear isoelectric focusing gradient of pH 3–10, followed by 12% SDS/PAGE. Gels were stained with silver. (**B**) Glycoproteins (1 mg) were first separated on a linear isoelectric focusing gradient of pH 4–7, followed by 12% SDS/PAGE. Gels were stained with Coomassie Brilliant Blue R-250. The circled spot was cut out of the gel, subjected to in-gel digestion with trypsin, and peptides were purified and analysed by ESI–Q–TOF–MS (Figure 4A).

of the glycopeptide yields a value of 804.44 Da, which is in good agreement with the theoretical value.

As mentioned above, transgenic or 'knockout' glycosyltransferase mutations in mice and other animals can be used to identify some of the protein targets of the glycosyltransferase being studied. One method of achieving this is to compare 2D–IEF–SDS/PAGE patterns of wild-type and mutant animals. This strategy can be applied to *C. elegans* by comparing wild-type worms with *gly-13* null worms.

A search of the Worm Proteome Database (WormPD; www.proteome.com/databases/index.html) with the search terms 'N-linked glycosylation', 'N-linked', 'N-glycosylation' and 'glycosylation' yielded 150 non-redundant 'hits'. Most of these are due either to the presence of the Asn-Xaa-Ser/Thr sequon or to the involvement of the protein in a glycosylation-related function (e.g. glycosyltransferases); in only a few cases (e.g. *ost-1*) was there direct evidence for N-glycan on the protein. Structural post-translational proteomics projects that target *C. elegans* glycans ('glycomics') by our group and others [110,121,122] will, hopefully, increase the number of experimentally established glycoproteins in the *C. elegans* database.

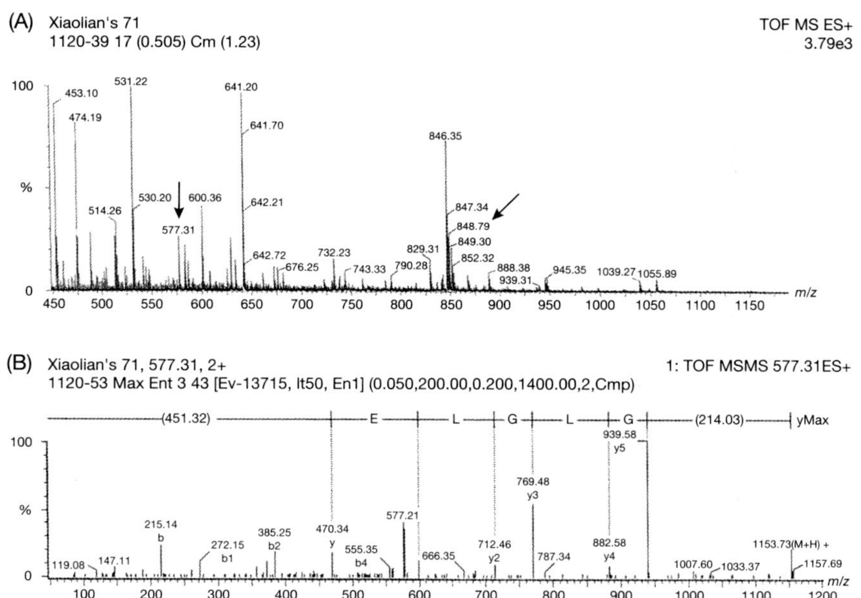

Figure 4 MS analysis of a *C. elegans* glycoprotein. (A) The tryptic digest of the putative glycoprotein detected by 2D–IEF–SDS/PAGE in Figure 3(B) was analysed by ESI–Q–TOF–MS. (**B**) The $\{M+2H\}^{2+}$ ion at *m/z* 577.31 (arrow in Figure 4A) was analysed by ESI–Q–TOF–MS/MS. The amino acid sequence tag Glu-Leu-Gly-Leu-Gly, ions at *m/z* 470.34 and 939.58 at the respective ends of Glu-Leu-Gly-Leu-Gly, and the mass of the neutral parent (1153.73) were submitted to the Peptide Search algorithm and yielded the sequence Thr-Ile-Gly-Leu-Gly-Leu-Glu-Pro-Ile-Ile-Lys. This allowed identification of hypothetical protein T13F3.6 with a molecular mass of 21 742 Da, which is consistent with the molecular mass observed by 2D–IEF–SDS/PAGE in Figure 3(B) (circled).

Step 2: screening for known *C. elegans* mutants

The next step in the procedure is to determine whether there are mutant worms corresponding to the glycoproteins identified in Step 1. If such mutants exist, they must have some sort of phenotype to be useful in our screening. When a potential glycoprotein target has been identified either by direct experimentation (Step 1, above) or by searching the WormPD, the entry for the relevant gene in the WormPD will contain information on mutant worm strains and their phenotypes. For example, the WormPD entry for the putative glycoprotein (CET-1, a member of the transforming growth factor-β superfamily) encoded by the gene *dbl-1* indicates that there are six mutant alleles known for this gene. CET-1 has a single Asn-Ala-Thr sequon, but it is not yet known whether this site is occupied by an N-glycan [123]. Worms with homozygous null mutations in *cet-1* have a readily detectable phenotype, i.e. shortened bodies and an abnormal male tail phenotype that resembles *sma* mutants (male sensory ray fusions), suggesting that *cet-1*, *sma-2*, *sma-3* and *sma-4* share a common pathway. CET-1 appears to affect body length in a dose-dependent manner. Heterozygotes for *cet-1* dis-

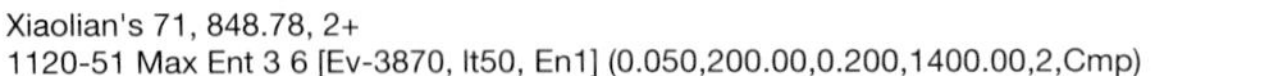

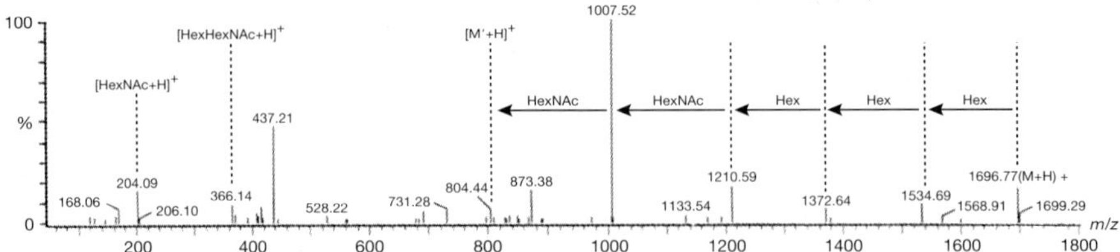

Figure 5 MS/MS analysis of a *C. elegans* glycopeptide. The {M+2H}$^{2+}$ ion at *m/z* 848.79 (arrow in Figure 4A) was analyzed by ESI–Q–TOF–MS/MS. The spectrum shows a glycopeptide with a molecular mass {M′+H}$^{1+}$ of 804.44 Da for the peptide moiety and a molecular mass {M′+H}$^{1+}$ of 892.33 Da for the glycan moiety. This identifies the glycan as Hex3HexNAc2 (see text).

played body lengths ranging between those observed for the null mutant and those for the wild-type. Injection of a plasmid construct encoding the wild-type *cet-1* genomic DNA was able to rescue the null mutant phenotype [123]. This observation is important for Step 3 of the screening.

Step 3: preparation of hypoglycosylating transgenic worms

Genes can be introduced into the *C. elegans* germ line of mutant strains by microinjection of DNA into the syncytial ovary of the hermaphrodite worm [98]. The ability to rescue the mutant worm phenotype with a wild-type transgene is essential to the screening. If rescue is successful, transgenic worms will be prepared with genes in which the Asn-Xaa-Ser/Thr sequon(s) have been mutated so that N-glycosylation cannot occur. If a sequon-mutated gene is unable to correct the phenotype, either completely or partially, the stage is set for studying the role of N-glycans of a specific glycoprotein on a specific step in worm development.

Failure to rescue a phenotype may result if the absence of glycans prevents proper protein folding with subsequent destruction of the protein by the ‘quality control’ mechanism of the cell. Quality control problems will be detected by insertion of an epitope-tagged (e.g. Myc) version of the sequon-mutated gene into the mutant worm followed by Western blot analysis on extracts of the resulting transgenic worm [98].

Conclusions

It is likely that at least 50% of the genes in any metazoan genome encode N-glycosylated proteins [74]. While methods for structural analysis of protein glycosylation on a proteomics scale are being developed and applied, there have been few, if any, attempts to develop a corresponding functional post-translational proteomics approach. The approach suggested in this chapter does not provide a comprehensive screening method, as many proteins in the proteome will undoubtedly not fulfil the criteria required for the screen. The method is therefore designed to complement other approaches towards understanding glycan functions.

The method requires a mutant organism with a genetic defect in the protein being studied and with a detectable phenotype. *C. elegans* has been chosen as an initial target organism to take advantage of the large number of known mutant strains in the database. Furthermore, scientists who study *C. elegans* have organized a Knockout Consortium (http://elegans.bcgsc.bc.ca/knockouts.html), which produces worms with null mutations on request and stocks a fairly large number of previously prepared null mutants. Other organisms such as *D. melanogaster* can also be studied by this approach.

Functional post-translational proteomics is inherently not a high-throughput method, but rather a general protein-by-protein approach to identify glycoproteins in the proteome that require their glycans for normal function. Our method uses normal development of the worm as a biological 'read-out' and should identify proteins essential for development that carry one or more asparagine-linked N-glycans necessary to this function. Although much of our previous work has focused on the role of hybrid and complex N-glycans in development, the proteomics approach targets all N-glycans. An analogous approach targeting only hybrid and complex N-glycans on individual proteins has not as yet been devised. When a glycoprotein has been positively identified by the screening, the specific biochemical and biological mechanisms of glycan function will require the kind of detailed study described above (see the section 'Function of specific glycans on specific glycoproteins').

The task of deciphering the mechanisms of action of glycosylation and other post translational modifications is certainly daunting; however, the exciting recent work on proteins such as Notch and dystroglycan (described above) indicate that the rewards will undoubtedly be worth the effort.

J.C.R. was a post-doctoral fellow at the University of New Hampshire supported by FAPESP fellowship 99/00487-8. H.S., A.M.S., J.W.C. and D.J.M. were supported by grants from the Canadian Institutes of Health Research and the Canadian Protein Engineering Network of Centres of Excellence. H.S. was also supported by the Mizutani Foundation for Glycoscience. V.N.R. was supported by grants from the National Institutes of Health.

References

1. Michalski, J.-C. (1996) in Glycoproteins and Disease, vol. 30 (Montreuil, J., Vliegenthart, J.F.G. and Schachter, H., eds), pp. 55–97, Elsevier, Amsterdam
2. Callahan, J.W. (1999) Biochim. Biophys. Acta **1455**, 85–103
3. Kornfeld, R., Bao, M., Brewer, K., Noll, C. and Canfield, W. (1999) J. Biol. Chem. **274**, 32778–32785
4. Raas-Rothschild, A., Cormier-Daire, V., Bao, M., Genin, E., Salomon, R., Brewer, K., Zeigler, M., Mandel, H., Toth, S., Roe, B. et al. (2000) J. Clin. Invest. **105**, 673–681
5. Kornfeld, S. (1990) Biochem. Soc. Trans. **18**, 367–374
6. Kresse, H. (1996) in Glycoproteins and Disease, vol. 30 (Montreuil, J., Vliegenthart, J.F.G. and Schachter, H., eds.), pp. 331–336, Elsevier, Amsterdam
7. Okajima, T., Fukumoto, S., Furukawa, K. and Urano, T. (1999) J. Biol. Chem. **274**, 28841–28844
8. Becker, D.J. and Lowe, J.B. (1999) Biochim. Biophys. Acta **1455**, 193–204

9. Jaeken, J., Matthijs, G., Carchon, H. and Van Schaftingen, E. (2001) in The Metabolic and Molecular Bases of Inherited Disease, vol. 1 (Scriver, C.R., Beaudet, A.L., Sly, W.S. and Valle, D., eds.), pp. 1601–1622, McGraw-Hill Medical Publishing, New York
10. Freeze, H.H. and Westphal, V. (2001) Biochimie **83**, 791–799
11. Schachter, H. (2001) J. Clin. Invest. **108**, 1579–1582
12. Schachter, H. (2001) Cell. Mol. Life Sci. **58**, 1085–1104
13. Metzler, M., Gertz, A., Sarkar, M., Schachter, H., Schrader, J.W. and Marth, J.D. (1994) EMBO J. **13**, 2056–2065
14. Ioffe, E. and Stanley, P. (1994) Proc. Natl. Acad. Sci. U.S.A. **91**, 728–732
15. Wang, Y., Tan, J., Sutton-Smith, M., Ditto, D., Panico, M., Campbell, R.M., Varki, N.M., Long, J.M., Jaeken, J., Levinson, S.R. et al. (2001) Glycobiology **11**, 874
16. Wang, Y., Tan, J., Sutton-Smith, M., Ditto, D., Panico, M., Campbell, R.M., Varki, N.M., Long, J.M., Jaeken, J., Levinson, S.R., Wynshaw-Boris, A., Morris, H.R., Le, D., Dell, A., Schachter, H. and Marth, J.D. (2001) Glycobiology, **11**, 1051–1070
17. Priatel, J.J., Sarkar, M., Schachter, H. and Marth, J.D. (1997) Glycobiology **7**, 45–56
18. Bhaumik, M., Sundaram, S. and Stanley, P. (1996) Glycobiology **6**, 720
19. Bhaumik, M., Harris, T., Sundaram, S., Johnson, L., Guttenplan, J., Rogler, C. and Stanley, P. (1998) Cancer Res. **58**, 2881–2887
20. Granovsky, M., Fata, J., Pawling, J., Muller, W.J., Khokha, R. and Dennis, J.W. (2000) Nat. Med. **6**, 306–312
21. Demetriou, M., Granovsky, M., Quaggin, S. and Dennis, J.W. (2001) Nature (London) **409**, 733–739
22. Furukawa, K. and Sato, T. (1999) Biochim. Biophys. Acta **1473**, 54–66
23. Lu, Q.X., Hasty, P. and Shur, B.D. (1997) Dev. Biol. **181**, 257–267
24. Asano, M., Furukawa, K., Kido, M., Matsumoto, S., Umesaki, Y., Kochibe, N. and Iwakura, Y. (1997) EMBO J. **16**, 1850–1857
25. Chui, D., Oh-Eda, M., Liao, Y.F., Panneerselvam, K., Lal, A., Marek, K.W., Freeze, H.H., Moremen, K.W., Fukuda, M.N. and Marth, J.D. (1997) Cell **90**, 157–167
26. Chui, D., Sellakumar, G., Green, R., Sutton-Smith, M., McQuistan, T., Marek, K., Morris, H., Dell, A. and Marth, J. (2001) Proc. Natl. Acad. Sci. U.S.A. **98**, 1142–1147
27. Maly, P., Thall, A.D., Petryniak, B., Rogers, G.E., Smith, P.L., Marks, R.M., Kelly, R.J., Gersten, K.M., Cheng, G.Y., Saunders, T.L. et al. (1996) Cell **86**, 643–653
28. Lowe, J.B. (2001) Cell **104**, 809–812
29. Homeister, J.W., Thall, A.D., Petryniak, B., Maly, P., Rogers, C.E., Smith, P.L., Kelly, R.J., Gersten, K.M., Askari, S.W., Cheng, G. et al. (2001) Immunity **15**, 115–126
30. Shafi, R., Iyer, S.P., Ellies, L.G., O'Donnell, N., Marek, K.W., Chui, D., Hart, G.W. and Marth, J.D. (2000) Proc. Natl. Acad. Sci. U.S.A. **97**, 5735–5739
31. Grewal, P.K., Holzfeind, P.J., Bittner, R.E. and Hewitt, J.E. (2001) Nat. Genet. **28**, 151–154
32. Hennet, T., Chui, D., Paulson, J.C. and Marth, J.D. (1998) Proc. Natl. Acad. Sci. U.S.A. **95**, 4504–4509
33. Ellies, L.G., Tsuboi, S., Petryniak, B., Lowe, J.B., Fukuda, M. and Marth, J.D. (1998) Immunity **9**, 881–890
34. Marek, K.W., Vijay, I.K. and Marth, J.D. (1999) Glycobiology **9**, 1263–1271
35. Priatel, J.J., Chui, D., Hiraoka, N., Simmons, C.J., Richardson, K.B., Page, D.M., Fukuda, M., Varki, N.M. and Marth, J.D. (2000) Immunity **12**, 273–283
36. Snapp, K.R., Heitzig, C.E., Ellies, L.G., Marth, J.D. and Kansas, G.S. (2001) Blood **97**, 3806–3811
37. Sperandio, M., Thatte, A., Foy, D., Ellies, L.G., Marth, J.D. and Ley, K. (2001) Blood **97**, 3812–3819
38. Sperandio, M., Forlow, S.B., Thatte, J., Ellies, L.G., Marth, J.D. and Ley, K. (2001) J. Immunol. **167**, 2268–2274

39. Yeh, J.C., Hiraoka, N., Petryniak, B., Nakayama, J., Ellies, L.G., Rabuka, D., Hindsgaul, O., Marth, J.D., Lowe, J.B. and Fukuda, M. (2001) Cell **105**, 957–969
40. Stanley, P., Narasimhan, S., Siminovitch, L. and Schachter, H. (1975) Proc. Natl. Acad. Sci. U.S.A. **72**, 3323–3327
41. Stanley, P., Caillibot, V. and Siminovitch, L. (1975) Cell **6**, 121–128
42. Gottlieb, C., Baenziger, J. and Kornfeld, S. (1975) J. Biol. Chem. **250**, 3303–3309
43. Vischer, P. and Hughes, R.C. (1981) Eur. J. Biochem. **117**, 275–284
44. Selleck, S.B. (2000) Trends Genet. **16**, 206–212
45. Selleck, S.B. (2001) Semin. Cell. Dev. Biol. **12**, 127–134
46. Herman, T., Hartwieg, E. and Horvitz, H.R. (1999) Proc. Natl. Acad. Sci. U.S.A. **96**, 968–973
47. Hirabayashi, J. (1999) Trends Glycosci. Glycotechnol. **11**, 211–212
48. Bulik, D.A., Wei, G., Toyoda, H., Kinoshita-Toyoda, A., Waldrip, W.R., Esko, J.D., Robbins, P.W. and Selleck, S.B. (2000) Proc. Natl. Acad. Sci. U.S.A. **97**, 10838–10843
49. Berninsone, P., Hwang, H.Y., Zemtseva, I., Horvitz, H.R. and Hirschberg, C.B. (2001) Proc. Natl. Acad. Sci. U.S.A. **98**, 3738–3743
50. Okajima, T., Yoshida, K., Kondo, T. and Furukawa, K. (1999) J. Biol. Chem. **274**, 22915–22918
51. Culetto, E. and Sattelle, D.B. (2000) Hum. Mol. Genet. **9**, 869–877
52. Kornfeld, R., Bao, M., Brewer, K., Noll, C. and Canfield, W.M. (1998) J. Biol. Chem. **273**, 23203–23210
53. Bao, M., Booth, J.L., Elmendorf, B.J. and Canfield, W.M. (1996) J. Biol. Chem. **271**, 31437–31445
54. Bao, M., Elmendorf, B.J., Booth, J.L., Drake, R.R. and Canfield, W.M. (1996) J. Biol. Chem. **271**, 31446–31451
55. Dennis, J.W., Granovsky, M. and Warren, C.E. (1999) Bioessays **21**, 412–421
56. McEver, R.P., Moore, K.L. and Cummings, R.D. (1995) J. Biol. Chem. **270**, 11025–11028
57. McEver, R.P. (1997) Glycoconjugate J. **14**, 585–591
58. Crottet, P., Kim, Y.J. and Varki, A. (1996) Glycobiology **6**, 191–208
59. Bruckner, K., Perez, L., Clausen, H. and Cohen, S. (2000) Nature (London) **406**, 411–415
60. Moloney, D.J., Panin, V.M., Johnston, S.H., Chen, J., Shao, L., Wilson, R., Wang, Y., Stanley, P., Irvine, K.D., Haltiwanger, R.S. and Vogt, T.F. (2000) Nature (London) **406**, 369–375
61. Wang, Y., Shao, L., Shi, S., Harris, R.J., Spellman, M.W., Stanley, P. and Haltiwanger, R.S. (2001) J. Biol. Chem. **276**, 40338–40345
62. Moloney, D.J., Shair, L.H., Lu, F.M., Xia, J., Locke, R., Matta, K.L. and Haltiwanger, R.S. (2000) J. Biol. Chem. **275**, 9604–9611
63. Chen, J., Moloney, D.J. and Stanley, P. (2001) Proc. Natl. Acad. Sci. U.S.A. **98**, 13716–13721
64. Holt, K.H., Crosbie, R.H., Venzke, D.P. and Campbell, K.P. (2000) FEBS Lett. **468**, 79–83
65. Williamson, R.A., Henry, M.D., Daniels, K.J., Hrstka, R.F., Lee, J.C., Sunada, Y., Ibraghimov-Beskrovnaya, O. and Campbell, K.P. (1997) Hum. Mol. Genet. **6**, 831–841
66. Chiba, A., Matsumura, K., Yamada, H., Inazu, T., Shimizu, T., Kusunoki, S., Kanazawa, I., Kobata, A. and Endo, T. (1997) J. Biol. Chem. **272**, 2156–2162
67. Sasaki, T., Yamada, H., Matsumura, K., Shimizu, T., Kobata, A. and Endo, T. (1998) Biochim. Biophys. Acta **1425**, 599–606
68. Endo, T. (1999) Biochim. Biophys. Acta **1473**, 237–246
69. Betel, D., Zhang, W. and Schachter, H. (2000) Glycobiology **10**, 1103–1104
70. Betel, D., Zhang, W. and Schachter, H. (2000) Glycoconj. J. **17**, 41
71. Zhang, W., Betel, D. and Schachter, H. (2001) Glycobiology **11**, 875
72. Zhang, W., Betel, D. and Schachter, H. (2002) Biochem. J. **361**, 153–162
73. Yoshida, A., Kobayashi, K., Manya, H., Taniguchi, K., Kano, H., Mizuno, M., Inazu, T., Mitsuhashi, H., Takahashi, S., Takeuchi, M. et al. (2001) Dev. Cell **1**, 717–724

74. Apweiler, R., Hermjakob, H. and Sharon, N. (1999) Biochim. Biophys. Acta **1473**, 4–8
75. Krishna, R.G. and Wold, F. (1993) Adv. Enzymol. Relat. Areas Mol. Biol. **67**, 265–298
76. Yan, S.C., Grinnell, B.W. and Wold, F. (1989) Trends Biochem. Sci. **14**, 264–268
77. Stanley, P. (1998) Trends Cell. Biol. **8**, 128–130
78. Stanley, P. and Ioffe, E. (1995) FASEB J. **9**, 1436–1444
79. Stanley, P., Raju, T.S. and Bhaumik, M. (1996) Glycobiology **6**, 695–699
80. Esko, J.D., Stewart, T.E. and Taylor, W.H. (1985) Proc. Natl. Acad. Sci. U.S.A. **82**, 3197–3201
81. Kingsley, D.M., Kozarsky, K.F., Hobbie, L. and Krieger, M. (1986) Cell **44**, 749–759
82. Stanley, P. (2000) in Molecular and Cellular Glycobiology, vol. 30 (Fukuda, M. and Hindsgaul, O., eds), pp. 169–198, Oxford University Press, Oxford
83. Verma, R., Annan, R.S., Huddleston, M.J., Carr, S.A., Reynard, G. and Deshaies, R.J. (1997) Science **278**, 455–460
84. Annan, R.S. and Carr, S.A. (1997) J. Protein Chem. **16**, 391–402
85. Wilm, M., Neubauer, G. and Mann, M. (1996) Anal. Chem. **68**, 527–533
86. Oda, Y., Nagasu, T. and Chait, B.T. (2001) Nat. Biotechnol. **19**, 379–382
87. Adamczyk, M., Gebler, J.C. and Wu, J. (2001) Rapid Commun. Mass Spectrom. **15**, 1481–1488
88. Goshe, M.B., Conrads, T.P., Panisko, E.A., Angell, N.H., Veenstra, T.D. and Smith, R.D. (2001) Anal. Chem. **73**, 2578–2586
89. Zhou, H., Watts, J.D. and Aebersold, R. (2001) Nat. Biotechnol. **19**, 375–378
90. Vliegenthart, J.F.G. and Casset, F. (1998) Curr. Opin. Struct. Biol. **8**, 565–571
91. Dell, A., Reason, A.J., Khoo, K.H., Panico, M., Mcdowell, R.A. and Morris, H.R. (1994) in Guide to Techniques in Glycobiology, vol. 230 (Lennarz, W.J. and Hart, G.W., eds), pp. 108–132, Academic Press, San Diego, CA
92. Reinhold, V.N., Reinhold, B.B. and Chan, S. (1996) in High Resolution Separation and Analysis of Biological Macromolecules, part B, vol. 271 (Karger, B.L. and Hancock, W.S., eds), pp. 377–402, Academic Press, San Diego, CA
93. Costello, C.E. (1997) Biophys. Chem. **68**, 173–188
94. Haslam, S.M., Morris, H.R. and Dell, A. (2001) Trends Parasitol. **17**, 231–235
95. Wood, W.B. (1988) The Nematode Caenorhabditis Elegans, Cold Spring Harbor Laboratory Press, Cold Spring Harbor, NY
96. Riddle, D.L., Blumenthal, T., Meyer, B.J. and Priess, J.R. (1997) C. elegans II, vol. 33, Cold Spring Harbor Laboratory Press, Cold Spring Harbor, NY
97. Epstein, H.F. and Shakes, D.C. (eds) (1995) Methods in Cell Biology, vol. 48 *Caenorhabditis elegans*: Modern Biological Analysis of an Organism, Academic Press, San Diego, CA
98. Chen, S.H., Zhou, S.H., Sarkar, M., Spence, A.M. and Schachter, H. (1999) J. Biol. Chem. **274**, 288–297
99. Chen, S., Spence, A.M. and Schachter, H. (2000) Glycobiology **10**, 1114
100. Chen, S., Spence, A.M. and Schachter, H. (2000) Glycoconj. J. **17**, 71
101. Wilson, I.B., Zeleny, R., Kolarich, D., Staudacher, E., Stroop, C.J., Kamerling, J.P. and Altmann, F. (2001) Glycobiology **11**, 261–274
102. Johnson, K.D. and Chrispeels, M.J. (1987) Plant Physiol. **84**, 1301–1308
103. Vitale, A. and Chrispeels, M.J. (1984) J. Cell Biol. **99**, 133–140
104. Sturm, A. (1995) in Glycoproteins, vol. 29a (Montreuil, J., Vliegenthart, J.F.G. and Schachter, H., eds), pp. 521–541, Elsevier, Amsterdam
105. Altmann, F., Schwihla, H., Staudacher, E., Glossl, J. and Marz, L. (1995) J. Biol. Chem. **270**, 17344–17349
106. März, L., Altmann, F., Staudacher, E. and Kubelka, V. (1995) in Glycoproteins, vol. 29a (Montreuil, J., Vliegenthart, J.F. G. and Schachter, H., eds.), pp. 543–563, Elsevier, Amsterdam

107. Rosa, J.C., Schachter, H., Reinhold, B. and Reinhold, V.N. (2001) Glycobiology **11**, 884–885
108. Reinhold, V.N., Reinhold, B.B. and Costello, C.E. (1995) Anal. Chem. **67**, 1772–1784
109. Sheeley, D.M. and Reinhold, V.N. (1998) Anal. Chem. **70**, 3053–3059
110. Altmann, F., Fabini, G., Ahorn, H. and Wilson, I.B. (2001) Biochimie **83**, 703–712
111. Tan, J., Chen, S., Spence, A.M. and Schachter, H. (2000) Glycobiology **10**, 1115
112. Warren, C.E., Krizus, A., Partridge, E., Lau, J.Y.-T., Roy, P.J., Zhang, L., Culotti, J.G. and Dennis, J.W. (1999) Glycobiology **9**, 1111
113. DeBose-Boyd, R.A., Nyame, A.K. and Cummings, R.D. (1998) Glycobiology **8**, 905–917
114. Zheng, Q., Van Die, I. and Cummings, R. (2000) Glycoconj. J. **17**, 28
115. Harnett, W., Houston, K.M., Tate, R., Garate, T., Apfel, H., Adam, R., Haslam, S.M., Panico, M., Paxton, T., Dell, A. et al. (1999) Mol. Biochem. Parasitol. **104**, 11–23
116. Haslam, S.M., Khoo, K.-H., Houston, K.M., Harnett, W., Morris, H.R. and Dell, A. (1997) Mol. Biochem. Parasitol. **85**, 53–66
117. Haslam, S.M., Houston, K.M., Harnett, W., Reason, A.J., Morris, H.R. and Dell, A. (1999) J. Biol. Chem. **274**, 20953–20960
118. Morelle, W., Haslam, S.M., Olivier, V., Appleton, J.A., Morris, H.R. and Dell, A. (2000) Glycobiology **10**, 941–950
119. Bagshaw, R.D., Callahan, J.W. and Mahuran, D.J. (2000) Anal. Biochem. **284**, 432–435
120. Borchers, C., Peter, J.F., Hall, M.C., Kunkel, T.A. and Tomer, K.B. (2000) Anal. Chem. **72**, 1163–1168
121. Hirabayashi, J. and Kasai, K. (2000) Trends Glycosci. Glycotechnol. **12**, 1–5
122. Hirabayashi, J., Arata, Y. and Kasai, K. (2001) Proteomics **1**, 295–303
123. Morita, K., Chow, K.L. and Ueno, N. (1999) Development **126**, 1337–1347
124. Tull, D., Miao, S.C., Withers, S.G. and Aebersold, R. (1995) Anal. Biochem. **224**, 509–514
125. Gygi, S.P., Rist, B., Gerber, S.A., Turecek, F., Gelb, M.H. and Aebersold, R. (1999) Nat. Biotechnol. **17**, 994–999
126. Smolka, M.B., Zhou, H., Purkayastha, S. and Aebersold, R. (2001) Anal. Biochem. **297**, 25–31

Biochem. Soc. Symp. **69**, 23–32
(Printed in Great Britain)

2

Comparative aspects of glycosyltransferases

Christelle Breton*, Helena Heissigerová*†, Charlotte Jeanneau*, Jitka Moravcová† and Anne Imberty*[1]

*Centre de Recherches sur les Macromolécules Végétales, Centre National de la Recherche Scientifique, BP 53, 38041 Grenoble cedex 9, France, and †Institute of Chemical Technology, Technicka 5, 166 28 Prague, Czech Republic

Abstract

Glycosyltransferases, the enzymes that build oligosaccharides and glycoconjugates, have received much interest in recent years owing to their biological functions and their potential uses in biotechnology. Despite the fact that many glycosyltransferases recognize similar donor or acceptor substrates, there is surprisingly limited sequence identity between different classes. On the one hand, the glycosyltransferases are found in a large number of families, by sequence-based classification. On the other hand, only two structural folds have been identified among the fewer than one dozen glycosyltransferases that have been crystallized at present. Detection of conserved motifs that have a direct role in the functional aspects of glycosyltransferases is one approach for identifying remote similarity. With the availability of more crystal structures, the use of the fold-recognition approach is also very promising.

Introduction

Glycosylation reactions are of great biological importance in both prokaryotes and eukaryotes, and require the co-ordinated action of a large number of enzymes, the glycosyltransferases (GTs). They catalyse the transfer of a sugar residue from an activated donor, usually a nucleotide sugar, to an acceptor that can be an oligosaccharide, a lipid or a protein. Recent developments in the molecular biology of GTs have revealed an unexpected diversity of these enzymes, which suggests that glycosylation reactions probably require the participation of several hundred genes.

[1]To whom correspondence should be addressed (e-mail anne.imberty@cermav.cnrs.fr).

Sequence analysis of GTs: motifs and patterns

GTs have been classified into 53 different families on the basis of amino-acid sequence similarities and stereochemistry of the reaction in the CAZY database (http://afmb.cnrs-mrs.fr/~cazy/CAZY/index.html); however, despite a lack of overall sequence identity, conserved peptide motifs have been detected in protein members belonging to different GT families. For example, a highly conserved peptide motif has been found in both eukaryotic and prokaryotic α2- and α6-fucosyltransferases (GT families 11, 23 and 37) [1,2], and in the peptide O-fucosyltransferase (C. Breton, unpublished work). Three conserved regions are also shared by N-acetylglucosaminyltransferase (GnT)-I and GnT-II, the key enzymes involved in the biosynthesis of complex N-glycans (GT families 13 and 16) [3]. The best example is the well-known Asp-Xaa-Asp motif, which has been identified in many different GT families that use various nucleotide sugars and acceptors [4,5]. This motif is present in both inverting and retaining enzymes, and it has been shown, in several crystal structures, to interact mainly with the phosphate groups of the nucleotide donor through the co-ordination of a metal cation. From crystal structures and mutagenesis experiments, it is clear that the conserved peptide motifs play a major role in enzyme function and their presence in distantly related proteins is indicative of common structural and catalytic features.

Comparative analysis of crystal structures of GTs

GTs present peculiar difficulties for expression, purification and crystallization and, as a result, they have been excluded for a while from the mainstream of the exploding number of crystal structures that have been solved. Since the first structure of bacteriophage T4 β-glucosyltransferase (BGT), an enzyme that adds glucose to the modified phage DNA [6], crystal structures have been obtained for eight other GTs. These structures are listed in Table 1. Although the number of structures available is still limited, very interesting information can be extracted because both prokaryotic and eukaryotic enzymes (including inverting and retaining enzymes) are present. Furthermore, in many cases, ligands (UDP, UDP-sugar, acceptor or Mn^{2+}) are present in the crystal, allowing study of the way that they interact with the protein.

All of the nine GTs with known structures belong to different sequence families. Unexpectedly, they appear to adopt only two different folds. In this chapter, they are named as the BGT and SpsA fold, with reference to the first structure solved in each case (see Table 1). The BGT-fold consists of two Rossmann-type domains that are connected by a flexible hinge and terminated by a series of long α-helices (Figure 1) . The similarity between the two domains is high enough to propose that they are the result of gene duplication [7]. Binding of the nucleotide sugar and of the acceptor is found in the crevice between the two domains, but little is known about the catalytic mechanism at the present time. The SpsA fold is characterized by an α/β/α sandwich that also shares similarity with the Rossmann fold. The central β-sheet is flanked by a smaller one, and the association of the two sheets creates the cleft for binding

Table 1 Crystal structures of GTs. Glc, glucose; Gal, galactose; GlcNAc, *N*-acetylglucosamine; Man, mannose; Xyl, xylose; NC, non-classified; GlcA, glucuronic acid.

GT	Organism	Donor	Acceptor	Link	Family	Fold	Reference
BGT	Phage T4	UDP-Glc	Modified DNA	β-	NC	BGT	[6,21,22]
SpsA	*Bacillus subtilis*	Unknown	Unknown		GT-2	SpsA	[23,24]
β4GalT1	Bovine	UDP-Gal	GlcNAc/Glc	β-4	GT-7	SpsA	[11,25,26]
MurG	*Escherichia coli*	UDP-GlcNAc	Peptidoglycan	β-	GT-28	BGT	[27]
GlcAT	Human	UDP-GlcA	Galβ1,4Xyl	β-3	GT-43	SpsA	[28]
GnT-I	Rabbit	UDP-GlcNAc	Manα1,3Man	β-2	GT-13	SpsA	[8]
α3GalT	Bovine	UDP-Gal	Galβ1,4GlcNAc	α-3	GT-6	SpsA	[10,25]
LgtC	*Neisseria meningitidis*	UDP-Gal	Galβ1,4Glc	α-4	GT-8	SpsA	[9]
GtfB	*Amycolatopsis orientalis*	UDP-Glc	Heptapeptide	β-	GT-1	BGT	[7]

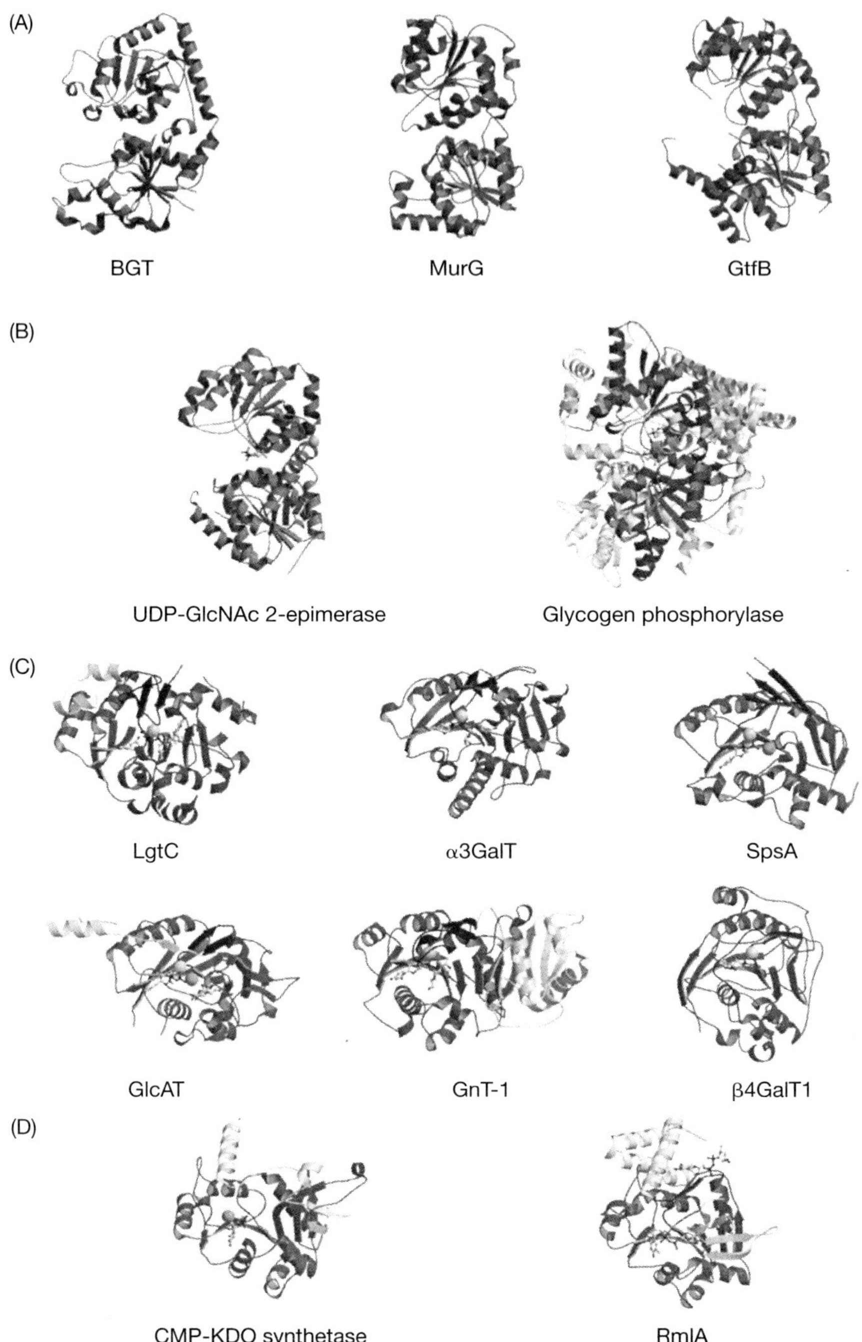

Figure 1 Ribbon representation of crystal structures of GTs and of some proteins sharing the same fold. (**A**) GTs with the BGT-fold. (**B**) Other proteins with the BGT fold. (**C**) GTs with the SpsA fold. (**D**) Selected other proteins with the SpsA fold. See Table 1 and Table 2 for references. GlcNAc, *N*-acetylglucosamine; KDO, 3-deoxy-manno-octulosonic acid.

the nucleotide sugar, the acceptor and the Mn^{2+} cation that have been found in all enzymes sharing this fold. For both GnT-I [8] and the galactosyltransferase LgtC [9], the structure has been obtained as a complex with the divalent cation and the nucleotide sugar, and the role of a flexible C-terminal loop that is able to lock the site has been demonstrated. Recent results suggest that large conformational changes occur upon binding of the nucleotide sugar donor [10,11].

Comparative studies with other proteins sharing the same fold

When looking for both the BGT and SpsA folds in the SCOP (Structural Classification of Proteins [12]) and FSSP [13] databases, only a limited number of other structures are identified, most of them being related to the glycosylation pathway. Figure 1 illustrates some of these similarities. UDP-N-acetylglucosamine 2-epimerase [14] adopts the same fold as BGT and binds the nucleotide sugar in the same region. More surprisingly, it has been demonstrated that BGT displays a strong structural resemblance to the core domains of glycogen phosphorylase and maltodextrin phosphorylases [15,16]. It is difficult to decide if this similarity is the result of a remarkable convergent evolution or of a very remote evolutionary relationship.

The enzymes that share the same fold as SpsA are listed in Table 2. They all bind nucleotide triphosphate, but catalyse two different reactions. On the one hand, CMP-*N*-acetylneuraminic acid (CMP-NeuAc) and CMP–3-deoxy-manno-octulosonic acid (CMP-KDO) synthetases catalyse the activation of acidic sugar by forming a monophosphate diester. On the other hand, the other enzymes create the nucleoside diphosphate diesters from the phosphorylated acceptor, which is generally a sugar. The two families of enzymes require the presence of Mg^{2+} for activity, although it appears that the location of this cation is not always equivalent to the one of Mn^{2+} in GTs.

Fold-recognition studies of GTs

The fact that GT families present little or no sequence identity and only two three-dimensional folds makes them very suitable proteins for the use of fold-recognition methods. In one approach, we analysed several representative sequences of each of the CAZY families, with the use of the 3D-PSSM web-based program [17] (www.bmm.icnet.uk/~3dpssm/). Families corresponding to integral membrane proteins with multiple transmembrane domains (22, 39, 48, 50 and 53), as well as those corresponding to phosphorylases and murein polymerases (36 and 51) were not considered. The results are listed in Table 3 for the families where no crystal structure is available. Of the 40 remaining families whose structures are unknown, 24 could be assigned a BGT or a SpsA fold with a high degree of probability and nine with a reasonable probability level (the latter ones being referred to as 'BGT-like' or 'SpsA-like'). Either of these two folds could be predicted for families 17 and 44 (including GnT-III, which is responsible for the biosynthesis of the bisecting β4-GlcNAc in

Table 2 Crystal structures of proteins adopting the SpsA fold that do not present GT activity. Me, 2-C-methylerythritol; GlcNAc-1-P, *N*-acetylglucosamine 1-phosphate; NeuAc, *N*-acetylneuraminic acid.

Enzyme	Donor	Acceptor	Product	Reference
CMP-KDO synthetase	CTP	KDO	CMP-KDO	[29]
CMP-NeuAc synthetase	CTP	NeuAc	CMP-NeuAc	[30]
GlmU	UTP	GlcNAc-1-P	UDP-GlcNAc	[31]
RmlA	dTTP	Glc-1-P	dTDP-Glc	[32]
CDP-Me synthase	CTP	2-C-methylerythritol-4-P	CDP-Me	[33]
MobA	GTP	Molybdopterin	Molybdopterin guanidine diphosphate	[17]

Table 3 Prediction of folds for GT families of unknown structure. GlcNAc-PI, *N*-acetylglucosamine–phosphatidylinositol; Man, mannose; GalNAc, *N*-acetylgalactosamine; ManNAcA, *N*-acetylmannuronic acid; Fuc, fucose; Xyl, xylose; Gal*f*, galactofuranose; LPS, lipopolysaccharide; OGT, O-GlcNAc transferase; Fuc4NAc, *N*-acetylfucosamine; GAG, glycosaminoglycan. With regard to the enzyme abbreviations, 'T' stands for transferase, and follows the abbreviated name of the monosaccharide being transferred.

Type of fold	GT family	Known functions
BGT	4	GlcNAc-PI synthase (PigA), LPS α-GlcNAcT, α-GlcT, α-ManT (Alg2), α-GalT, galactolipid synthase, sucrose synthase
	5	Glycogen synthase, starch synthase
	9	Bacterial LPS-GlcNAcT, heptosyltransferase
	19	Lipid A disaccharide synthase (LpxB)
	20	α,α-Trehalose-phosphate synthase
	30	CMP-KDO transferase
	33	β-ManT (Alg1)
	38	Bacterial α2,8-polysialyltransferase
	41	O-linked β-GlcNAcT (OGT)
	47	N-term domain of bi-functional heparan sulphate synthase (EXT family)
	56	TDP-Fuc4NAc:lipid II Fuc4NAc transferase
'BGT-like'	3	Glycogen synthase
	10	α3/4-Fucosyltransferases
	11	Bacterial and animal α2-fucosyltransferases
	23	α6-Fucosyltransferases
	29	Eukaryotic sialyltransferases
	37	Xyloglucan α2-fucosyltransferase
SpsA	12	β4-GalNAcT (GM2/GD2 synthase)
	16	β2-GlcNAcT (GnT-II)
	21	Ceramide b-GlcT
	24	Glycoprotein α3-GlcT
	25	LPS β4-GalT (LgtB, LgtE)
	27	Polypeptide α-GalNAcT
	31	β3-GalT, β3-GlcNAcT (Fringe)
	34	α2-GalT (Mnn10, Gma12), galactomannan α6-GalT
	40	Bacterial Gal*f*T
	45	Bacterial α-GlcNAcT
	46	Bacterial sequences of unknown function
	47	C-terminal domain of bi-functional heparan sulphate synthase (EXT family)
	55	Bacterial α-ManT
'SpsA-like'	32	α6-ManT, α4GlcNAcT/α4-GalT
	49	β3-GlcNAcT, C-terminal domain of LARGE protein
	54	β4-GlcNAcT (GnT-IV, GnT-VI)

(contd.) ☞

Table 3 (contd.)

Type of fold	GT family	Known functions
Rossmann domain	14	β6-GlcNAcT (C2GnT), β-XylT (GAG synthesis)
	15	Yeast α2-ManT
	18	β6-GlcNAcT (GnT-V)
	26	Bacterial UDP-ManNAcA transferase, β-GlcT
	42	Bacterial α2,3/α2,8-sialyltransferase
	52	Bacterial α2,3-sialyltransferase

N-glycans). Families 14, 15, 18, 26, 42 and 52 are predicted to contain a Rosmmann-like fold, but they could not be assigned unambiguously to the SpsA or BGT families. For some bi-functional enzymes, such as EXT proteins (e.g. GT-47), a BGT fold was predicted for the N-terminal domain and a SpsA fold for the C-terminal domain. The LARGE protein is predicted to be a bi-functional enzyme displaying two separate SpsA domains, with the N-terminal domain showing significant similarities with the GT-8 protein sequences (LgtC family), whereas the C-terminal domain is similar to a β3-GlcNAcT (GT-49). Using a different fold-recognition program, Wrabl and Grishin made rather similar predictions for the occurrence of a BGT fold in several CAZY families [18].

Figure 2 illustrates the prediction that can be performed for some of the enzymes involved in the biosynthesis of N-glycans. The mannosyltransferases Alg1 (GT-33) and Alg2 (GT-4), which participate to the synthesis of the dolichol phosphate oligosaccharide precursor, are predicted to adopt the BGT fold. Among the Golgi-localized enzymes, GnT-I and β4-GalT1 have been crystallized and are known to adopt the SpsA fold. The same structure is predicted for

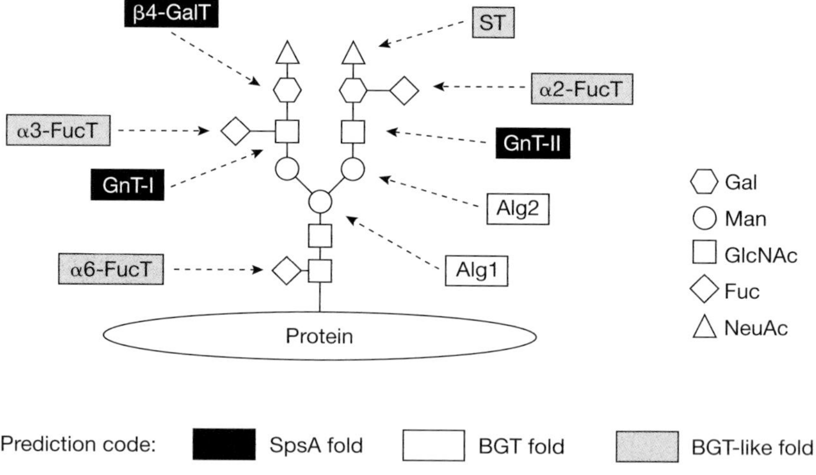

Figure 2 Schematic representation of a complex-type N-glycan together with labelling of the GTs involved in its biosynthesis. The enzymes are shaded according to their predicted fold.

GnT-II, the threading prediction being supported by the existence of common sequence motifs between GnT-I and GnT-II [3]. At present, there are no experimental structural data for sialyltransferases (STs) and fucosyltransferases (FTs). The 3D-PSSM approach allows us to predict that all theses enzymes are likely to adopt the BGT fold. Our previous threading work, which used a different program, arrived at the same prediction for fucosyltransferases [2,19] but it should be noted that another team arrived at a different model [20]. This clearly illustrates that more structural data are needed on these enzymes of high biological and pharmaceutical interest.

Perspectives: combination of fold recognition and motif search for identification of GT activities

Annotation of newly determined genomic sequences is a bottleneck in modern biology. When looking at the genomes of bacteria or parasites, the number of annotated GTs is notably smaller than what would be necessary to build the highly complex polysaccharides or oligosaccharides present at the cell surface or in the cell wall. The use of the fold-recognition approach, associated with the search for functional motifs, may in the future allow the identification of 'putative' GTs that could be the target for functional studies.

References

1. Breton, C., Oriol, R. and Imberty, A. (1998) Glycobiology **8**, 87–94
2. Chazalet, V., Uehara, K., Geremia, R.A. and Breton, C. (2001) J. Bacteriol. **183**, 7067–7075
3. Breton, C., Mucha, J. and Jeanneau, C. (2001) Biochimie **83**, 713–718
4. Breton, C., Bettler, E., Joziasse, D.H., Geremia, R.A. and Imberty, A. (1998) J. Biochem. **123**, 1000–1009
5. Wiggins, C.A. and Munro, S. (1998) Proc. Natl. Acad. Sci. U.S.A. **95**, 7945–7950
6. Vrielink, A., Rüger, W., Driessen, H.P.C. and Freemont, P.S. (1994) EMBO J. **13**, 3413–3422
7. Mulichak, A.M., Losey, H.C., Walsh, C.T. and Garavito, R.M. (2001) Structure **9**, 547–557
8. Ünligil, U.M., Zhou, S., Yuwaraj, S., Sarkar, M., Schachter, H. and Rini, J.M. (2000) EMBO J. **19**, 5269–5280
9. Persson, K., Ly, H.D., Dieckelmann, M., Wakarchuk, W.W., Withers, S.G. and Strynadka, N.C.J. (2001) Nat. Struct. Biol. **8**, 166–175
10. Boix, E., Swaminathan, G.J., Zhang, Y., Natesh, R., Brew, K. and Acharya, K.R. (2001) J. Biol. Chem. **276**, 48608–48614
11. Ramakrishnan, B. and Qasba, P.K. (2001) J. Mol. Biol. **310**, 205–218
12. Murzin, A.G., Brenner, S.E., Hubbard, T. and Chothia, C. (1995) J. Mol. Biol. **247**, 536–540
13. Holm, L. and Sander, C. (1996) Science **273**, 595–602
14. Campbell, R.E., Mosimann, S.C., Tanner, M.E. and Strynadka, N.C. (2000) Biochemistry **39**, 14993–15001
15. Artymiuk, P.J., Rice, D.W., Poirrette, A.R. and Wilett, P. (1995) Nat. Struct. Biol. **2**, 117–120
16. Holm, L. and Sander, C. (1995) EMBO J. **14**, 1287–1293
17. Lake, M.W., Temple, C.A., Rajagopalan, K.V. and Schindelin, H. (2000) J. Biol. Chem. **275**, 40211–40217
18. Wrabl, J. and Grishin, N. (2001) J. Mol. Biol. **314**, 365–374
19. Breton, C., Oriol, R. and Imberty, A. (1996) Glycobiology **6**, vii–xii

20. de Vries, T., Yen, T.Y., Joshi, R.K., Storm, J., van Den Eijnden, D.H., Knegtel, R.M., Bunschoten, H., Joziasse, D.H. and Macher, B.A. (2001) Glycobiology **11**, 423–432
21. Moréra, S., Imberty, A., Aschke-Sonnenbron, Rüger, W. and Freemont, P.S. (1999) J. Mol. Biol. **292**, 717–730
22. Moréra, S., Lariviere, L., Kurzeck, J., Aschke-Sonnenborn, U., Freemont, P.S., Janin, J. and Ruger, W. (2001) J. Mol. Biol. **311**, 569-577
23. Charnock, S.J. and Davies, G.J. (1999) Biochemistry **38**, 6380–6385
24. Tarbouriech, N., Charnock, S.J. and Davies, G.J. (2001) J. Mol. Biol. **314**, 655–661
25. Gastinel, L.N., Cambillau, C. and Bourne, Y. (1999) EMBO J. **18**, 3546–3557
26. Ramakrishnan, B., Shah, P.S. and Qasba, P.K. (2001) J. Biol. Chem. **276**, 37665–37671
27. Ha, S., Walker, D., Shi, Y. and Walker, S. (2000) Protein Sci. **9**, 1045–1052
28. Pedersen, L.C., Tsuchida, K., Kitagawa, H., Sugahara, K., Darden, T.A. and Negishi, M. (2000) J. Biol. Chem. **275**, 34580–34585
29. Jelakovic, S. and Schulz, G.E. (2001) J. Mol. Biol. **312**, 143–155
30. Mosimann, S.C., Gilbert, M., Dombroswki, D., To, R., Wakarchuk, W. and Strynadka, N.C. (2001) J. Biol. Chem. **276**, 8190–8196
31. Brown, K., Pompeo, F., Dixon, S., Mengin-Lecreulx, D., Cambillau, C. and Bourne, Y. (1999) EMBO J. **18**, 4096–4107
32. Blankenfeldt, W., Asuncion, M., Lam, J.S. and Naismith, J.H. (2000) EMBO J. **19**, 6652–6663
33. Richard, S.B., Bowman, M.E., Kwiatkowski, W., Kang, I., Chow, C., Lillo, A.M., Cane, D.E. and Noel, J.P. (2001) Nat. Struct. Biol. **8**, 641–648

Biochem. Soc. Symp. **69**, 33–46
(Printed in Great Britain)

3

Glycosyltransferases and glycan structures contributing to the adhesive activities of L-, E- and P-selectin counter-receptors

John B. Lowe[1]

Howard Hughes Medical Institute, Department of Pathology, University of Michigan Medical School, 1150 West Medical Center Drive, Ann Arbor, MI 48109-0650, U.S.A.

Abstract

In mammals, leucocytes of the adaptive and innate immune systems must move from their sites of origin to sites of maturation, or to where they are deployed against the invasion of pathogens. The vascular tree serves as the primary throughfare by which leucocytes move to these various destinations. Adhesion must be established between the leucocyte and the endothelial cells that line the vascular tree to enable leucocytes to escape the vascular compartment and then contribute to extravascular immune processes. A major fraction of these leucocyte–endothelial-cell adhesive events initiate with, and require interactions between, the selectin family of cell adhesion molecules and their glycoconjugate counter-receptors. This article will review the structures of the glycan components of these counter-receptors, and the glycosyltransferases that control their expression.

Introduction

The use of molecular cloning during the last 15 years has begun to provide a glimpse at the protein and corresponding DNA sequences of the enzymes that control glycan structure and function in mammals. Generally speaking, glycan structural diversity is regulated during developmental processes and shows lineage specificity in the adult animal. This diversity is determined largely by transcriptional control of the glycosyltransferases that are responsible for glycoconjugate synthesis [1]. A consideration of the glycan

[1]E-mail johnlowe@umich.edu

structures themselves, as well as emerging information about expression patterns of the various glycosyltransferases involved in their synthesis, allows one to assign glycosyltransferases and their cognate glycosidic bonds to two general types (Figure 1). The first type is classified based on glycan moieties, and the corresponding glycosyltransferases that are constitutively expressed in most or even all cell types in the mammalian organism. These glycan structures represent, to some extent, a platform upon which the second type of glycosyltransferases operate. Members of this type are often expressed with developmental and lineage specificity, and can thus yield glycan structures that are found only in some cell types, and/or only at certain stages of development in mammals. Both types of glycosyltransferases and the corresponding glycans are essential to the elaboration of glycoprotein counter-receptors for the selectin family of cell adhesion molecules, and in the control of selectin-dependent leucocyte-trafficking events in mammals [2]. This chapter reviews the current understanding of the glycan structures that contribute to selectin counter-receptor activity, with an emphasis on the glycosyltransferase genes that control the synthesis of these oligosaccharides, and on genomic approaches that have helped to identify and characterize these genes.

As reviewed elsewhere, the selectin family of cell adhesion molecules comprises three molecules: E-selectin, P-selectin and L-selectin [2,3]. These are type 1 transmembrane proteins, each of which has an N-terminal carbohydrate recognition domain. E-selectin expression is generally restricted to vascular

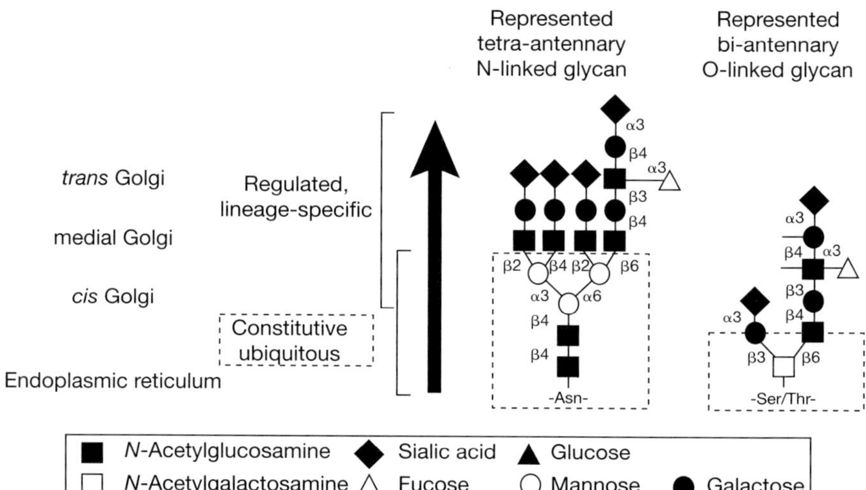

Figure 1 Constitutive and regulated glycosyltransferases. Prototypical asparagine-linked glycans, serine/threonine-linked glycans and lipid-linked glycans are shown. Enzymes that account for the synthesis of the glycosidic bonds closest to the plasma membrane are more likely to be constitutively expressed in many or even all tissues. In contrast, temporal regulation of expression during development and lineage specificity of restriction are more typical of glycosidic bonds, and the corresponding glycosyltransferases at the more distal reaches of these types of mammalian glycans. See text for additional discussion.

endothelial cells, is induced under the influence of cytokines like tumour necrosis factor α and interleukin 1β, and recognizes sialylated, fucosylated glycoprotein and glycolipid counter-receptors on leucocytes. Inflammatory stimuli induce expression of P-selectin by vascular endothelial cells; acutely by liberation from preformed P-selectin molecules stored in Weibel–Palade bodies, and subacutely after transcriptional induction. P-selectin is also elaborated by platelets after platelet activation. Leucocyte counter-receptors for P-selectin are, like E-selectin counter-receptors, sialylated and fucosylated glycoconjugates [2].

Recruitment of neutrophils, monocytes, eosinophils and specific lymphocyte subsets to extravascular inflammatory sites is, in most instances, essentially dependent upon E-selectin- and P-selectin-dependent adhesion interaction with the glycoconjugate counter-receptors on leucocytes. In the context of shear forces within the vasculature, E-selectin- and P-selectin-dependent leucocyte adhesion captures free-flowing leucocytes, and mediates leucocyte rolling under shear. Activational events caused by cytokines from the vascular wall activate the rolling leucocytes, yielding inside–out signalling events that lead to leucocyte integrin activation, firm adhesion to the endothelial cell lining of the vascular wall via integrin-dependent adhesion, followed by transleucocyte migration. L-selectin, which is expressed by most types of blood leucocytes, also contributes to this process, either by displaying counter-receptor activity for E-selectin, or by recognizing glycoconjugate counter-receptors expressed by activated vascular endothelia. This multi-step paradigm has been reviewed in [2,3].

A distinct leucocyte trafficking process, termed lymphocyte homing, is, by contrast, mediated exclusively by L-selectin-dependent adhesion. Lymphocyte homing is the process by which T- and B-lymphocytes are recruited from the blood into the parenchyma of lymph nodes and Peyer's patches. Lymphocyte-borne L-selectin mediates lymphocyte rolling on specialized postcapillary endothelial vessels — high-endothelial venules (HEVs) — in these secondary lymphoid organs via interactions with specific sialylated, fucosylated, sulphated mucin-type L-selectin counter-receptors that are expressed by the endothelial cells of the HEVs. As with E-selectin- and P-selectin-dependent leucocyte recruitment and inflammation, L-selectin-dependent lymphocyte homing is characterized by lymphocyte rolling under shear, followed by integrin activation, integrin-dependent firm adhesion and lymphocyte transmigration to the parenchyma of the node. This paradigm, in the context of lymphocyte homing, has been reviewed in [3,4].

L-selectin ligands required for lymphocyte homing

The molecular details of L-selectin glycoconjugate counter-receptors on HEVs, and the corresponding glycosyltransferase genes that are necessary for their synthesis, are understood in substantial detail. These are discussed first to provide a conceptual framework for the subsequent discussion of E-selectin and P-selectin counter-receptor biosynthesis in leucocytes, for which there are many strong parallels, but where there is a less extensive and precise understanding of the enzymes and structures involved.

Initial clues as to the glyconjugate nature of L-selectin counter-receptors derived from studies by Rosen's group, which demonstrated a requirement for sialylation and sulphation in the interaction between L-selectin and its counter-receptors (reviewed in [4]). A role for fucosylation in L-selectin counter-receptor activity was suggested by the identification of fucose as a component of L-selectin counter-receptors [5], by a requirement for α1-3fucosylation in E-selectin and P-selectin counter-receptor activity *in vitro* [2,6] and by a requirement for α1-3fucosylation in lymphocyte homing and L-selectin ligand activity *in vivo* [6]. Armed with the knowledge that L-selectin counter-receptor activity required sulphation, sialylation and α1-3fucosylation, glycan structure analyses were initiated to define, at a molecular level, the structures of the glycans required for L-selectin counter-receptor activity [7–9]. These analyses provided evidence for the existence of one or more specific sialylated, sulphated, fucosylated, glycan 'capping' groups associated with L-selectin counter-receptors purified from mouse HEVs (Figure 2). These studies represented an important starting point for identifying the genes required for the synthesis of these glycans, the order of their assembly and their specific contributions to L-selectin-dependent cell adhesion.

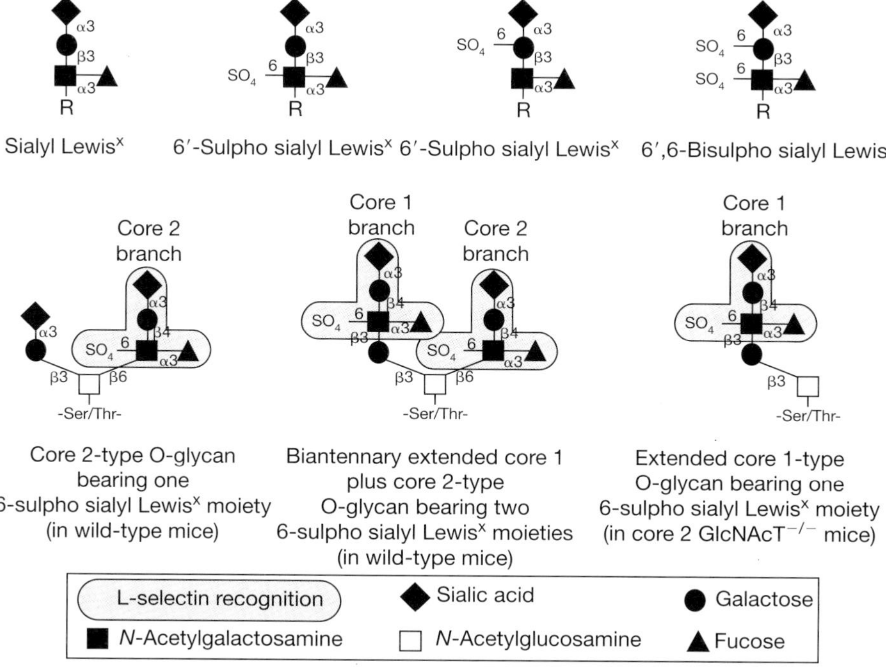

Figure 2 Glycans that covalently modify glycosylation-dpendent cell adhesion molecule 1 (GlyCAM-1) and other glycoproteins that exhibit L-selectin counter-receptor activity

Sulphotransferases in L-selectin ligand synthesis

These capping groups were characterized by the presence of sulphate at the C-6 position of the galactose residue within the tetrasaccharide capping group, and sulphate at the C-6 position of the *N*-acetylglucosamine (GlcNAc) residue within this moiety. As discussed elsewhere, biochemical and genetic evidence support a biosynthetic scheme for the generation of these proposed capping groups, starting with sulphation of glycan precursors, followed by α2-3sialylation, and then α1-3fucosylation [4,10], yet the nature of the sulphotransferase(s) required for construction of these capping groups had not been defined. Molecular cloning efforts had previously identified a sulphotransferase gene responsible for 6-O-galactose sulphation in keratin sulphate biosynthesis [11], and a related sulphotransferase gene encoding a 6-O-GalNAc sulphotransferase involved in chondroitin sulphate biosynthesis [12]. Biochemical studies with these enzymes indicate that both enzymes can add sulphate at the C-6 position of the galactose moiety on α2-3sialyl-*N*-acetyl-lactosamine with a low efficiency [11,12]. These considerations, together with weak primary sequence similarities among the catalytic domains of several sulphotransferases, prompted two groups to use segments of these genes to search human and mouse sequence databases for related, candidate sulphotransferases corresponding to those involved in L-selectin counter-receptor sulphation. In one instance, these efforts identified a cDNA encoding a protein termed L-selectin ligand sulphotransferase (LSST), a predicted type 2 transmembrane protein that is typical of mammalian glycosyltransferases [13]. In another instance, the human version of this sequence, which was called high endothelial cell (HEC)-GlcNAc-6-sulphotransferase, was cloned [14]. Both groups also studied a second enzyme exhibiting GlcNAc-6-sulphotransferase activity. These enzymes are representative of a family of galactose/N-acetylgalactosamine/GlcNAc 6-O-sulphotransferases (GSTs). These enzymes, their substrate specificities, nomenclature and chromosomal localization have been reviewed [4,15].

In vitro studies have implied that one member of this family, termed GST-3 (formerly known as LSST or HEC-GlcNAc-6-sulphotransferase), is likely to have the most direct relevance to L-selectin ligand carbohydrate sulphation [13,14,16]. These studies show that GST-3 transcripts localize primarily to HECs, which is consistent with a role in the expression of L-selectin ligands by HEVs. In cultured cells that have been transfected with the mucin-type glycoprotein CD34, together with a core 2 GlcNAc transferase (Core-2-GlcNAcT) and the α1-3fucosyltransferase FucT-VII, L-selectin ligand activity is clearly GST-3-dependent, as assessed by flow cytometry experiments using an L-selectin–Ig chimaera, or by cell adhesion assays under shear. Biochemical analyses indicate clearly that this enzyme catalyses the transfer of sulphate from the sulphation donor adenosine 3′-phosphate 5′-phosphosulphate to the 6-hydroxyl of the GlcNAc moiety on core-2-type O-glycans within the context of the sialyl Lewis^x tetrasaccharide. Considered together, observations from both groups provide strong evidence that GST-3 is responsible for synthesis of the 6-sulpho sialyl Lewis^x determinant on L-selectin counter-receptors that include CD34, glycosylation-dependent cell adhesion molecule 1 (GlyCAM-1) and mucosal addressin cell adhesion molecule 1 (MAdCAM-1) [13,14,16].

A role for this enzyme in the control of L-selectin ligand activity on HEVs has been demonstrated recently through the construction and analysis of a strain of mice with an induced null mutation in the GST-3 locus [17]. These studies showed that deletion of GST-3 yields a substantial reduction in L-selectin ligand activity on peripheral node HEVs, including loss of binding of an L-selectin–Ig chimaera to the luminal face of the HEV. The loss of binding correlates with a substantial decrease in homing of lymphocytes to the peripheral nodes, as well as a decrease in the steady state level of lymphocytes that can be recovered from the peripheral nodes. Considered together, these results assign GST-3 to a pivotal role in the elaboration of L-selectin ligand activities displayed by peripheral node HEVs, and that decorate the HEV-expressed L-selectin counter-receptor molecules CD34, GlyCAM-1 and other mucin-type glycoproteins. Interestingly, the GST-3 null mice appear to retain residual L-selectin ligand activity expressed by the luminal aspects of their HEVs in peripheral nodes, indicating the existence of GST-3-independent L-selectin ligand activities. It remains to be determined if these residual GST-3-independent activities are sulphation-dependent, and the nature of the glycan structures that account for these residual activities remains to be defined.

Core-2 and Core-1 O-glycans in L-selectin ligand activity

As is implied above, glycan structural analyses provide evidence for a contribution to the synthesis of glycans associated with L-selectin ligand activity by a β1-6GlcNAcT enzyme activity that is generically referred to as Core-2-GlcNAcT. This enzyme creates a core 2 branch on O-glycans, and, as noted above, GST-3 activity is apparently optimal with core 2 branched O-glycans. Furthermore, as reviewed elsewhere [3], core 2 O-glycans had been assigned an essential role in P-selectin counter-receptor activity in experiments where such activity was reconstituted in transfected cell lines *in vitro*. To assess the functional relevance of this enzyme in selectin ligand activity *in vivo*, a gene targeting approach has been used to construct a strain of mice that is deficient in a Core-2-GlcNAcT [18]. Subsequent to the creation and analysis of this mouse, additional molecular cloning work identified a family of three distinct Core-2-GlcNAcT loci, which corresponded to enzymes that are now termed Core-2-GlcNAcT-I, Core-2-GlcNAcT-II and Core-2-GlcNAcT-III [19,20]. The Core-2-GlcNAcT locus that was disabled in these knockout mouse studies corresponds to Core-2-GlcNAcT-I. Analysis of lymphocyte homing in the Core-2-GlcNAcT-I null mice reveals essentially normal lymphocyte homing to peripheral nodes, as well as to mesenteric nodes and Peyer's patches, and virtually normal accumulation of lymphocytes in the nodes [18]. A modest decrease in L-selectin ligands on peripheral node HEVs was observed, using immunohistochemistry and an L-selectin–IgM chimaera, in the null animals. These observations suggested that deletion of Core-2-GlcNAcT-I yields a subtle quantitative or qualitative decrease in L-selectin ligand reactivity that is not accompanied by a physiologically apparent lymphocyte homing defect.

The virtually normal L-selectin ligand activity in Core-2-GlcNAcT-I null animals could, in principle, be ascribed to expression of other Core-2-GlcNAcTs, or to L-selectin ligand activities that do not require Core-2-GlcNAcT activity. These alternative possibilities have been distinguished subsequently by analysis of the glycans expressed by the HEVs in Core-2-GlcNAcT-I null mice [21]. These studies disclose an absence of any core 2-type O-glycan structures among sulphated, sialylated radiolabelled O-glycans that were isolated from the HEVs of Core-2-GlcNAcT-I null mice, and thus exclude the possibility that other Core-2-GlcNAcT activities account for, more or less, normal L-selectin-dependent homing activities in Core-2-GlcNAcT-I null mice. Importantly, structural analyses of glycans expressed by the HEVs of Core-2-GlcNAcT-I null mice disclosed the existence of extended core 1-type O-linked glycans associated with HEV-derived mucin-type glycoproteins, and bearing the 6-sulpho sialyl Lewis^{x} structure. These observations predicted the existence of a corresponding Core-1-β3GlcNAcT.

Previous work demonstrated modest sequence similarities between a β3GlcNAc-T and three β1-3-galactosyltransferases, each of which exhibited some limited aspects corresponding to the requisite Core-1-β3GlcNAcT activity [22]. Searches of DNA sequence databases were initiated to identify candidate cDNAs corresponding to such an activity [21]. These efforts yielded a human DNA sequence whose corresponding transcripts localize to HEVs, as was expected for its possible role in the synthesis of L-selectin ligands. Its full-length cDNA is predicted to encode a protein with a type 2 transmembrane topology. A soluble fusion protein derivative of this candidate Core-1-β3GlcNAcT was observed to add GlcNAc to a low-molecular-mass core 1 glycan analogue (Galβ1-3GalNAcα1-R) to form the core 1 branch, and was also observed to add GlcNAc to a low-molecular-mass core 2 acceptor substrate [Galβ1-3(GlcNAcβ1-6)GalNAcα1-R]. Furthermore, reconstitution experiments using Chinese hamster ovary cells that express GlyCAM-1 or CD34, GST-3 and Fuc-TVII, with either Core-1-β3GlcNAcT, Core-2-GlcNAcT-I or both enzymes, disclose that expression of Core-1-β3GlcNAcT or of Core-2-GlcNAcT-I enhances L-selectin ligand activity of the transfected Chinese hamster ovary cells, as assessed by lymphocyte rolling under shear. By contrast, co-expression of Core-1-β3GlcNAcT plus Core-2-GlcNAcT-I yields an additive, and possibly synergistic, contribution to L-selectin ligand activity in this *in vitro* system. Glycan structural analyses to develop a molecular correlate for these adhesion studies identify bi-antennary O-glycans containing 6-sulpho sialyl Lewis^{x} on both the core 1 extension and on the core 2 branch. These observations indicate that both enzymes have important roles in wild-type HEVs. This work also suggests that, among the known 6-sulpho sialyl Lewis^{x}-type glycans expressed by peripheral node HEVs, the bi-antennary-type 6-sulpho sialyl Lewis^{x} structures may provide the most prominent contribution to L-selectin ligand activity, relative to 6-sulpho sialyl Lewis^{x}-substituted mono-antennary-type structures. Manipulation of the extended core 1 structures, via targeted deletion of the locus in mice, will help to define more precisely the functional relevance of Core-1-β3GlcNAcT in the expression of L-selectin ligands in HEVs.

Work in this paper also identified the molecular nature of the glycan epitope recognized by the MECA-79 monoclonal antibody. The MECA-79 antibody has been known for many years to recognize the luminal surface of peripheral node HEVs [23], and has been used widely to inhibit attachment of lymphocytes to HEVs *in vitro* and *in vivo* [23] and to isolate L-selectin ligands biochemically [24]; it has also been known to recognize sulphated O-glycans [25]. Nevertheless, the precise structural requirements for its recognition were not clear. Yeh et al. [21] firmly established that the MECA-79 epitope corresponds to the 6-sulphated, extended core 1 glycan Galβ1-4[6-sulpho]GlcNAcβ1-3Galβ1-3GalNAcα1.

Fucosylation in the control of L-selectin ligand activity

A contribution by α1-3fucosylation to the function of L-selectin counter-receptors was suggested by studies published in the early 1990s that identified a requirement for this modification for E-selectin and P-selectin counter-receptors (reviewed in [2]), and by the strong primary sequence similarity shared by the carbohydrate-recognition domains of E-, P- and L-selectins. Evidence to support the requirement for this modification for L-selectin ligand activities *in vivo* derives from the study of mice that were homozygous for a null mutation at an α1-3fucosyltransferase locus that is expressed in HEVs [26] and termed FucT-VII [6]. HEVs of the peripheral nodes in these mice do not bind an L-selectin–IgM chimaera and there is an approx. 80–90% reduction in the number of lymphocytes that can be recovered from the peripheral nodes of these animals, with a quantitatively corresponding defect in lymphocyte homing. These observations, together with *in vitro* studies indicating that FucT-VII can utilize 6-O-sulphated α2-3sialylated lactosamine precursors, implied that FucT-VII contributes to the synthesis of the 6-sulpho sialyl Lewis^x moiety discussed above.

The contribution of FucT-VII to L-selectin ligand synthesis in cultured cells is well established. In these studies, L-selectin ligands have been reconstituted through transfection with FucT-VII, glycoprotein ‘scaffolds’ like GlyCAM-1 or CD34, sulphotransferase cDNAs (including GST-3) and Core-2-GlcNAcT-I, while taking advantage of endogenous α2-3sialyltransferase activities [13,14,16,21,27]. These studies, and observations made in FucT-VII null mice, indicated that this enzyme is a primary participant in controlling L-selectin counter-receptor activity. Nevertheless, the peripheral nodes in FucT-VII null mice are not small, as had been observed in mice homozygous for null alleles at the L-selectin locus [27], and the nodes from FucT-VII null mice had some residual L-selectin-dependent lymphocyte homing activity. These considerations implied either that fucosylation-independent L-selectin ligand activities exist, or that another fucosyltransferase contributed to the residual activities. This uncertainty has been resolved with analysis of mice with deficiency of both FucT-VII and a second α1-3fucosyltransferase termed FucT-IV [10]. These mice have extremely small peripheral lymph nodes, akin to those observed in L-selectin null mice. The residual lymphocyte homing observed in the FucT-VII null mice is essentially abolished in the FucT-IV$^{-/-}$/FucT-VII$^{-/-}$ double-deficient mice, and their peripheral nodes are virtually devoid of naïve T-lymphocytes [28], a category of T-cells that require L-selectin for homing to peripheral lymph nodes [29]. By contrast, lymphocyte homing and peripheral node phenotypes in mice with a FucT-IV deficiency alone were essentially indistinguishable from those of wild-type mice.

Shear-dependent cell adhesion correlates for each of these phenotypes have been developed using *in vitro* shear-dependent cell adhesion assays. These assays take advantage of the fact that GlyCAM-1 is a membrane-associated, HEV-derived protein that is released into the circulation, and can be purified from the serum of mice [30]. Wild-type GlyCAM-1 and GlyCAM-1 from FucT-IV null mice, FucT-VII null mice and FucT-IV$^{-/-}$/FucT-VII$^{-/-}$ mice were purified from the serum of these mice and were attached to the bottom surface of a parallel plate flow chamber at two different densities. Wild-type mouse lymphocytes were then passed over this L-selectin counter-receptor-coated surface under physiologically relevant shear forces, and lymphocyte tethering events were captured by videomicroscopy and quantitated. These experiments demonstrated similar tethering activities of GlyCAM-1 isolated from wild-type and FucT-IV null mice however, they demonstrated a substantial decrease in tethering activity using GlyCAM-1 from FucT-VII null mice, and no detectable tethering using GlyCAM-1 isolated from the serum of the FucT-IV$^{-/-}$/FucT-VII$^{-/-}$ mice [10]. Considered together, these observations indicate that FucT-VII plays a much more prominent role than FucT-IV in the fucosylation of functionally relevant L-selectin counter-receptors. A possible role of FucT-IV is revealed only when FucT-VII is absent. Enzymic correlates for these observations, developed using low-molecular-mass precursors for some of the molecules in depicted Figure 2 suggest that FucT-VII contributes to the synthesis of 6-sulpho sialyl Lewis^x determinants and sialyl Lewis^x determinants, whereas FucT-IV may contribute to the synthesis of non-sialylated isomers of these molecules. Structurally related molecules are known to support L-selectin-dependent adhesion *in vitro* [31]; however, it remains to be determined if non-sialylated, 6-O-sulphated, α1-3fucosylated molecules are found in HEVs *in vivo*. It is also not known if such molecules are present in amounts that make meaningful contributions to L-selectin ligand activity and the lymphocyte homing process, and whether they may be increased in the absence of FucT-VII. Structural analysis of GlyCAM-1-associated glycans from the three different fucosyltransferase strains is in progress in order to address these issues. Future work to more precisely define the structures and biosynthesis of L-selectin ligands will most likely focus on defining glycan structures in mice with induced null mutations in new sulphotransferase and α2-3sialyltransferase loci that may participate in the synthesis of such glycans.

Glycosylation in the control of P- and E-selectin activity

Strong similarities exist between the glycosylation pathways that determine L-selectin counter-receptor activities and those required for E-selectin and P-selectin counter-receptor function, although it is safe to say that at present, there is less definitive glycan structural information in this regard. Radiochemical labelling procedures using HEVs in organ culture have helped to define glycan structures that are relevant to L-selectin ligands, whereas these procedures are not as effectively applied to leucocytes like blood neutrophils, whose glycan synthetic programme is considered to be nearly quiescent.

As noted above, early work assigned a primary role to sialyl Lewis^x-type glycans in E-selectin counter-receptor activity. Subsequent work identified E-selectin counter-receptor activity on some glycolipids [32], as well as specific glycoproteins on neutrophils, monocytes, T-lymphocyte subsets and eosinophils (reviewed in [3]). E-selectin counter-receptors on leucocytes are that presumed to or demonstrated to have sialyl Lewis^x-type glycans, include L-selectin itself [33] (in humans, at least) [34], N-glycans on ESL-1 [35] and a mucin-type glycoprotein termed PSGL-1 (P-selectin glycoprotein ligand 1; discussed in [36]). By contrast, P-selectin counter-receptor activity on leucocytes has been ascribed largely to PSGL-1 [3,36,37].

Reconstitution of P-selectin ligand activity in cultured cell lines requires PSGL-1, as well as α2-3sialylation, α1-3fucosylation, Core-2-GlcNAcT and protein tyrosine sulphation [38–40]. A detailed analysis of the O-glycan structures borne by PSGL-1 in the human promyelocytic leukaemia cell line HL-60 identifies a small set of core 2-type O-glycans bearing the sialyl Lewis^x determinant [41] (Figure 3). These include a monofucosylated core 2-type O-glycan with a sialyl Lewis^x capping group, and a multiply fucosylated poly-lactosamine-containing core 2 O-glycan also with a sialyl Lewis^x capping group. More recent studies indicate that a tyrosine-sulphated peptide derived from the N-terminal segment of PSGL-1, when decorated with the monofucosylated core 2-type O-glycan, binds to P-selectin with an affinity equivalent to native, leucocyte-derived PSGL-1, whereas the identical tyrosine-sulphated peptide, when decorated with the multiply fucosylated isomer, is not an effective P-selectin [42]. The function of the multiply fucosylated isomer, if it has any, remains to be defined, and the stoichiometry of modification of each relevant serine/threonine residue and its contribution to binding effectiveness must be determined. It also remains to be seen whether the structures identified

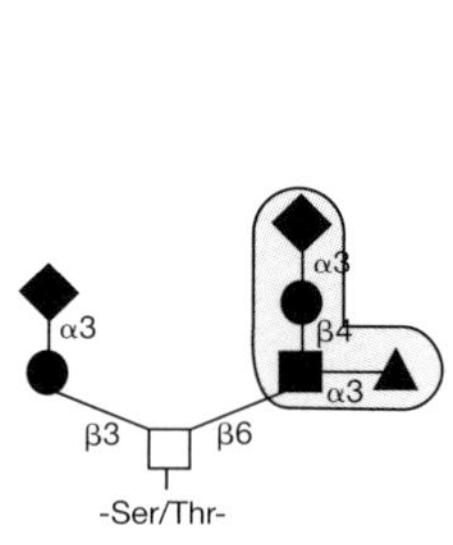

Core 2-type O-glycan bearing one sialyl Lewis^x moiety

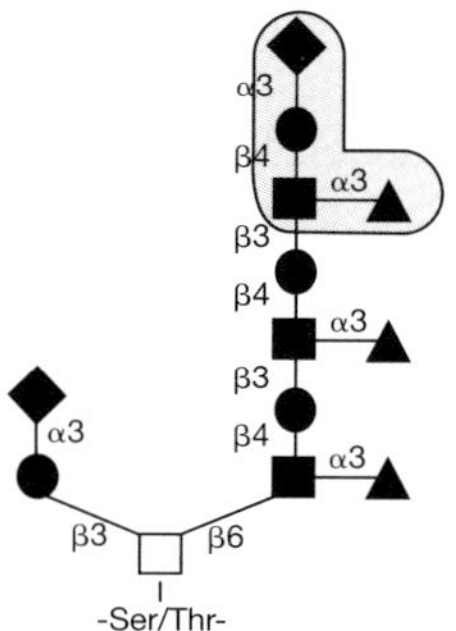

Core 2-type O-glycan bearing a terminal sialyl Lewis^x moiety and two internal α1-3fucosylations

Figure 3 Fucosylated glycans that covalently modify PSGL-1 on HL-60 cells

on the HL-60-derived PSGL-1 are representative of those on human neutrophils and other leucocytes, or on mouse leucocytes, where, as discussed below, glycosyltransferase gene deletion studies have helped to define the nature of the enzymes that direct the synthesis of E-selectin and P-selectin counter-receptor activities.

The Core-2-GlcNAcT-I null mice discussed above exhibit profound deficiency of neutrophil P-selectin ligand activity, and P-selectin-dependent cell adhesion [18]. These deficits are similar to the those observed in PSGL-1 null mice [36,37]. By contrast, a role for Core-2-GlcNAcT-I in leucocyte E-selectin ligand activity is less clear. Depending on the nature of the approach used to study contributions by Core-2-GlcNAcT-I and/or PSGL-1 to E-selectin ligand activity, substantial or subtle defects are observed, using either Core-2-GlcNAcT-I null mice [18,43,44] or PSGL-1 null mice [36,37]. Although these apparent discrepancies remain to be resolved, it seems likely that Core-2-GlcNAcT-I and/or PSGL-1 will be found to contribute substantially less to E-selectin counter-receptor activity than to that of P-selectin counter-receptor. It is interesting that Core-2-GlcNAcT-I is also implicated in the elaboration of L-selectin ligands expressed by inflammed vascular endothelium [45]. The nature of the corresponding glycans remains to be defined.

Analysis of mice with targeted deletions of the FucT-IV and FucT-VII loci indicate that the α1-3fucosyltransferase loci required for the synthesis of the glycans necessary for HEV-borne L-selectin ligand activity are identical to those required for P-selectin and E-selectin ligand activity. The α1-3fucosyltransferase FucT-VII is expressed at high levels in bone marrow, the source of myeloid lineage cells that are typically E-selectin-ligand- and P-selectin-ligand-positive [26]. In mice that are homozygous for an induced null mutation in the FucT-VII locus, a profound diminution in E-selectin and P-selectin counter-receptor activities is observed on neutrophils and monocytes [6,10]. Residual E-selectin and P-selectin ligand activity remains, however, as assessed in selectin-dependent inflammation models, and in shear-dependent flow chamber-based assays of neutrophil adhesion to recombinant E- and P-selectins. This residual activity is clearly FucT-IV-dependent, since the binding of neutrophils purified from FucT-IV$^{-/-}$/FucT-VII$^{-/-}$ mice to E- or P-selectin under shear cannot be detected *in vitro* or *in vivo*, and these mice are essentially unable to recruit neutrophils to extravascular sites in E-selectin- and P-selectin-dependent inflammatory models. By contrast, FucT-IV deficiency, where the FucT-VII locus is intact, yields neutrophils whose E-selectin and P-selectin counter-receptor activities are essentially equivalent to those of wild-type neutrophils, when assayed in shear-dependent adhesion *in vitro*, or in E-selectin- and P-selectin-dependent inflammatory models. However, subtle defects in the rolling velocities of FucT-IV null neutrophils are observed *in vivo*, using intravital microscopy [46]. These observations indicate that FucT-IV contributes to neutrophil E-selectin and P-selectin ligand activities, even when FucT-VII is present.

The FucT-VII locus is also expressed by subsets of T-lymphocytes that bind to E-selectins and P-selectins, and that use E-selectin and P-selectin to migrate to sites of cutaneous inflammation [2,28]. When isolated from FucT-

VII null mice, these cells, termed memory/effector cells of the Th1 and Tc1 phenotypes [28], neither bind to selectin *in vitro* nor migrate to inflammed skin *in vivo* [28]. In contrast, FucT-IV-deficient Th1- and Tc1-cells exhibit essentially normal E-selectin and P-selectin binding and *in vivo* trafficking properties. FucT-VII is thus clearly required for E-selectin and P-selectin counter-receptor activities on such cells, whereas Fuc-TIV is apparently dispensable. It remains to be seen if the small amounts of memory/effector lymphocyte trafficking observed in FucT-VII null mice is enabled by FucT-IV-dependent synthesis of E- and P-selectin counter-receptor activities.

Molecular correlates for these observations have yet be defined, and require structural analysis of the glycans borne by PSGL-1 on neutrophils, memory/effector lymphocytes and other leucocytes. The construction and analysis of Core-1-β3GlcNAcT null mice will help determine a role for Core-1-β3GlcNAcT in leucocyte E- and P-selectin ligand activities.

John Lowe is an Investigator of the Howard Hughes Medical Institute. This work was supported in part by grants from the National Institutes of Health.

References

1. Lowe, J.D. (2001) Cell **104**, 809–812
2. Vestweber, D. and Blanks, J.E. (1999) Physiol. Rev. **79**, 181–213
3. Butcher, E.C. and Picker, L.J. (1996) Science **272**, 60–66
4. Hemmerich, S. and Rosen, S.D. (2002) Glycobiology **10**, 849–856
5. Crommie, D. and Rosen, S.D. (1995) J. Biol. Chem. **270**, 22614–22624
6. Maly, .P, Thall, A., Petryniak, B., Rogers, C.E., Smith, P.L., Marks, R.M., Kelly, R.J., Gersten, K.M., Cheng, G., Saunders, T.L. et al. (1996) Cell **86**, 643–653
7. Hemmerich, S., Leffler, H. and Rosen, S.D. (1995) J. Biol. Chem. **270**, 12035–12047
8. Hemmerich, S. and Rosen, S.D. (1994) Biochemistry **33**, 4830–4835
9. Hemmerich, S., Bertozzi, C.R., Leffler, H. and Rosen, S.D. (1994) Biochemistry **33**, 4820–4829
10. Homeister, J.H., Thall, A., Petryniak, B., Maly, P., Rogers, C.E., Smith, P.L., Kelly, R.J., Gersten, K.M., Misra, A., Cheng, G. et al. (2001) Immunity **15**, 115–126
11. Torii, T., Fukuta, M. and Habuchi, O. (2000) Glycobiology **10**, 203–211
12. Habuchi, O., Suzuki, Y. and Fukuta, M. (1997) Glycobiology **7**, 405–412
13. Hiraoka, N., Tsuboi, S., Suzuki, M., Petryniak, B., Nakayama, J., Izawa, D., Tanaka, T., Miyasaka, M., Lowe, J.B. and Fukuda, M. (1999) Immunity **11**, 79–89
14. Bistrup, A., Bhakta, S., Lee, J.K., Belov, Y.Y., Gunn, M.D., Zuo, F.-R., Huang, C.-C., Kannagi, R., Rosen, S.D. and Hemmerich, S. (1999) J. Cell Biol. **145**, 899–910
15. Hemmerich, S., Lee, J.K., Bhakta, S., Bistrup, A., Ruddle, N.R. and Rosen, S.D. (2001) Glycobiology **11**, 75–87
16. Tangemann, K., Bistrup, A., Hemmerich, S. and Rosen, S.D. (1999). J. Exp. Med. **190**, 935–942
17. Hemmerich, S., Bistrup, A., Singer, M.S., van Zante, A., Lee, J.K., Tsay, D., Peter, M., Carminati, J.L., Brennan, T.J., Carver-Moore, K. et al. (2001) Immunity **15**, 237–247
18. Ellies, L.G., Tsuboi, S., Petryniak, B., Lowe, J.B., Fukuda, M. and Marth, J.D. (1998) Immunity **9**, 881–890
19. Yeh, J.C., Ong, E. and Fukuda, M. (1999) J. Biol. Chem. **274**, 3215–3221
20. Schwientek, T., Yeh, J.C., Levery, S.B., Keck, B., Merkx, G., van Kessel, A.G., Fukuda, M. and Clausen, H. (2000) J. Biol. Chem. **275**, 11106–11113

21. Yeh, J.-C., Hiraoka, N., Petryniak, B., Nakayama, J., Ellies, L.G., Rabuka, D., Hindsgaul, O., Marth, J.D., Lowe, J.B. and Fukuda, M. (2001) Cell **105**, 957–969
22. Zhou, D., Dinter, A., Gutierrez Gallego, R., Kamerling, J.P., Vliegenthart, J.F., Berger, E.G. and Hennet, T. (1999) Proc. Natl. Acad. Sci. U.S.A. **96**, 406–411
23. Streeter, P.R., Rouse, B.T. and Butcher, E.C. (1988) J. Cell. Biol. **107**, 1853–1862
24. Clark, R.A., Fuhlbrigge, R.C. and Springer, T.A. (1998) J. Cell. Biol. **140**, 721–731
25. Bruehl, R.E., Bertozzi, C.R. and Rosen, S.D. (2000) J. Biol. Chem. **275**, 32642–32648
26. Smith, P.L., Gersten, K.M., Petryniak, B., Kelly, R.J., Roger, C., Natsuka, Y., Alford, 3rd, J.A., Scheidegger, E.P., Natsuka, S. and Lowe, J.B. (1996) J. Biol. Chem. **1271**, 8250–8259
27. Kimura, N., Mitsuoka, C., Kanamori, A., Hiraiwa, N., Uchimura, K., Muramatsu, T., Tamatani, T., Kansas, G.S. and Kannagi, R. (1999) Proc. Natl. Acad. Sci. U.S.A. **96**, 4530–4535
28. Smithson, G., Rogers, C.E., Smith, P.L., Scheidegger, E.P., Petryniak, B., Myers, J.T., Kim, D.S., Homeister, J.W. and Lowe, J.B. (2001) J. Exp. Med. **194**, 601–614
29. Bradley, L.M., Watson, S.R. and Swain, S.L. (1994) J. Exp. Med. **180**, 2401–2406
30. Singer, M.S. and Rosen, S.D. (1996) J. Immunol. Methods **196**, 153–161
31. Galustian, C., Lubineau, A., le Narvor, C., Kiso, M., Brown, G. and Feizi, T. (1999) J. Biol. Chem. **274**, 18213–18217
32. Stroud, M.R., Handa, K., Salyan, M.E.K., Ito, K., Levery, S.B., Hakomori, S., Reinhold, B.B. and Reinhold, V.N. (1996). Biochemistry **35**, 770–778
33. Picker, L.J., Warnock, R.A., Burns, A.R., Doerschuk, C.M., Berg, E.L. and Butcher, E.C. (1991) Cell **66**, 921–933
34. Steegmaier, M., Levinovitz, A., Isenmann, S., Borges, E., Lenter, M., Kocher, H.P. and Vestweber, D. (1995) Nature (London) **373**, 615–620
35. Zöllner O., Lenter, M.C., Blanks J.E., Borges, E., Steegmaier, M., Zerwes, H.-G. and Vestweber, D. (1997) J. Cell Biol. **136**, 707–716
36. Xia, L., Sperandio, M., Yago, T., McDaniel, J.M., Cummings, R.D., Pearson-White, S., Ley, K. and McEver, R.P. (2002) J. Clin. Invest. **109**, 939–950
37. Yang, J., Hirata, T., Croce, K., Merrill-Skoloff, G., Tchernychev, B., Williams, E., Flaumenhaft, R., Furie, B.C. and Furie, B. (1999) J. Exp. Med. **190**, 1769–1782
38. Sako, D., Chang, X.J., Barone, K.M., Vachino, G., White, H.M., Shaw, G. and Veldman, G.M. (1993) Cell **75**, 1179–1186
39. Li, F., Wilkins, P.P., Crawley, S., Weinstein, J., Cummings, R.D. and McEver, R.P. (1996) J. Biol. Chem. **271**, 3255–3264
40. Wilkins, P.P., Moore, K.L., McEver, R.P. and Cummings, R.D. (1995) J. Biol. Chem. **270**, 22677–22680
41. Wilkins, P.P., McEver, R.P. and Cummings, R.D. (1996) J. Biol. Chem. **271**, 18732–18742
42. Leppänen, A., White, S.P., Helin, J., McEver, R.P. and Cummings, R.D. (2000) J. Biol. Chem. **275**, 39569–39578
43. Snapp, K.R., Heitzig, C.E., Ellies, L.G., Marth, J.D. and Kansas, G.S. (2001) Blood **97**, 3806–3811
44. Sperandio, M., Thatte, A., Foy, D., Ellies, L.G., Marth, J.D. and Ley, K. (2001) Blood **97**, 3812–3819
45. Sperandio, M., Forlow, S.B., Thatte, J., Ellies, L.G., Marth, J.D. and Ley, K. (2001) J. Immunol. **167**, 2268–2274
46. Weninger, W., Ulfman, L.H., Cheng, G., Souchkova, N., Quackenbush, E.J., Lowe, J.B. and von Andrian, U.H. (2000) Immunity **12**, 665–676

Biochem. Soc. Symp. **69**, 47–57
(Printed in Great Britain)

4

New insights into heparan sulphate biosynthesis from the study of mutant mice

Catherine L.R. Merry[1] and John T. Gallagher

Cancer Research UK Department of Medical Oncology, Christie Research Centre, Christie Hospital NHS Trust, Wilmslow Road, Manchester M20 4BX, U.K.

Abstract

Heparan sulphate (HS) is an essential co-receptor for a number of growth factors, morphogens and adhesion proteins. The biosynthetic modifications involved in the generation of a mature HS chain may determine the strength and outcome of HS–ligand interactions. These modifications are catalysed by a complex family of enzymes, some of which occur as multiple gene products. Various mutant mice have now been generated, which lack the function of isolated components of the HS biosynthetic pathway. In this discussion, we outline the key findings of these studies, and use them to put into context our own work concerning the structure of the HS generated by the *Hs2st*$^{-/-}$ mice.

Introduction

Recent advances in the cloning of many components of the heparan sulphate (HS) biosynthetic pathway have significantly increased our understanding of this complex process. Within this discussion we hope to outline how these studies have prompted a re-evaluation of the regulation of HS synthesis, and provided new insights into the varied functions of HS *in vivo*. Unlike the more extensively studied N-linked glycoforms, HS is an exclusively linear polysaccharide; its considerable structural heterogeneity is generated primarily by the pattern of modified sugars present within this chain [1,2]. HS consists of repeating disaccharide units of hexuronic acid (HexA) and glucosamine (GlcN) with typically between 50 and 200 units per chain, covalently linked to protein via a GlcA-Gal-Gal-Xyl tetrasaccharide, the latter having an O-glycosidic linkage to a serine residue. Two isomers of HexA, glucuronic acid (GlcA) and iduronic acid (IdoA), occur throughout the chain. GlcN residues

[1]To whom correspondence should be addressed (e-mail cmerry@picr.man.ac.uk).

can either be N-acetylated (*N*-acetylglucosamine, GlcNAc), N-sulphated (*N*-sulphylglucosamine, GlcNS) or, very rarely, unmodified. All of these variants can become O-sulphated at various positions; C-2 of IdoA and GlcA, and C-6 and C-3 of GlcN. Structurally, HS is similar to heparin, and is generated by the same biosynthetic pathway [2]; however, HS has a lower proportion of N-sulphated GlcN residues (approx. 50% in HS compared with approx. 100% in heparin) and a much-reduced number of O-sulphate additions [3]. In addition, sulphation in HS is clustered in domains, called sulphated- or S-domains, in contrast with the uniform sulphation found in heparin. The biological distribution of heparin is restricted almost exclusively to mature mast cells [4], although it is also present on a subset of neural precursors [5], while HS is a near-ubiquitous component of cell surfaces and the extracellular matrix, and as such is a key mediator of the interaction of cells with their local microenvironment [6].

HS proteoglycans (HSPGs) have been found on all adherent cells that have been investigated to date [7]. They are connected to the plasma membrane by a glycosylphosphatidylinositol anchor (a characteristic of the glypican family of HSPGs [8]) or by a transmembrane region in the core protein (a characteristic of the syndecan family [9]) and have the potential to occupy a variety of membrane niches, and to project for some distance from the cell surface [6]. In addition, HSPGs are essential components of the extracellular matrix and include multi-domain proteins such as perlecan, agrin and collagen XVIII [10].

HSPGs play key roles in many developmental processes [11,12]. They are obligate co-receptors for influential signalling molecules [13], and have other essential functions in forming morphogen gradients [14], protecting proteins from proteolytic degradation [15] and in defining cell identity [16]. During the last few years, novel experiments have been designed to take advantage of the cloning of a variety of HSPG core proteins. In various systems (notably *Caenhorabditis elegans*, *Drosophila* and mice), gene knockouts and loss-of-function studies have highlighted many previously unknown activities of HSPGs and enhanced our understanding of cellular and molecular processes in which HSPGs are known to be important (reviewed in [12,17–20]). In many systems, however, the biological functions of HSPGs are governed predominantly (but not exclusively) by their HS chains. It is via these chains that the majority of protein interactions take place, with a large and ever-growing list of growth factors, cytokines and extracellular matrix proteins known to have a strong affinity for HS. In agreement with this principle, previous studies have shown that the HS expressed by a particular cell/tissue is determined by the cell source itself rather than the core protein to which that HS is attached [21]. Therefore, although HS composition varies between different tissues [16,22,23], the same core proteins may well be present in these locations. Because of this, studies of single HSPG core-protein knockout animals will yield aberrant phenotypes on those rare occasions when a single HSPG is required for a particular interaction, be this because of an additional involvement of the core protein itself, or the role of that protein in the correct spatio-temporal positioning of the attached HS chains. Examples of this are perlecan in basement membrane integrity [24], syndecan 4 in wound healing [25,26] and glypican 3 in the regulation of cell growth [27]. The role of the HS chains themselves can be assessed more directly by generating mutants that lack components of the HS biosynthetic pathway.

Our ideas on HS biosynthesis are derived largely from studies on heparin produced by a cell-free system prepared from microsomal fractions of a mouse mastocytoma [28]. In particular, this controlled system enabled Lindahl and co-workers to dissect out most of the fundamental stages of HS/heparin synthesis (for a review and further references see [2]). Other methods also proved useful in elucidating the specificities of various enzymes of the biosynthetic machinery, such as the chemical mutagenesis of Chinese hamster ovary (CHO) cell lines which led to the isolation of clones deficient in chain synthesis or sulphation (reviewed in [29]). In combination with these studies, improved techniques for the analysis of HS domain structure and, more recently, sequence, have helped to generate libraries of 'permitted' sequences, which guided the definition of the specificities of the various biosynthetic enzymes [22,30,31]. The culmination of these studies resulted in the description of a pathway in which a succession of enzymes acted in a sequential manner, with each enzyme in sequence generating a potential substrate for the next [1] (Figure 1). Some of these were found to be dual-purpose enzymes, such as the HS polymerase (HS-pol) that catalyses the alternate addition of GlcA and GlcNAc residues to the nascent HS chain [32]. The next enzyme in the pathway, the N-deacetylase/N-sulphotransferase (NDST) also catalyses two events, the de-N-acetylation and re-N-sulphation of selected GlcNAc residues [33,34]. After this, a C-5 epimerase catalyses the conversion of some GlcA residues into IdoA. Many of these IdoA residues (and some GlcA resides) then become substrates for the 2-O-sulphotransferase (2-OST). Additional O-sulphation then occurs at C-6 of (predominantly) GlcNS residues, along with some GlcNAc sugars, catalysed by the 6-O-sulphotransferase (6-OST). Finally, a 3-O-sulphotransferase (3-OST) acts on isolated GlcN residues, typically modifying only a few sugars per chain [2]. These polymer-modifying enzymes were shown to be interdependent, and to operate in a concerted, but localized, manner along the chain, generating areas of high N- and O-sulphation alternating with regions of low modification in which N-acetylated disaccharides are the major structural unit [1,2].

Although the foregoing pathway suggested a mechanism for the polymer modification observed in HS and heparin, it was difficult to understand how it would be able to generate the domain structure and controlled molecular diversity known to exist in HS extracted from various cells and tissues. The recent advance in the cloning of many of the components of this pathway has brought us much closer to understanding these complex issues. One of the most significant discoveries was that many of the enzymes exist as multiple gene products, or as splice variants [2] (Figure 2). Within the mouse, for example, two genes encoding HS-pols have been identified so far (*EXT1* [35] and *EXT2* [36]; the murine counterparts of the human hereditary multiple exostoses genes), as well as four *NDST* genes [4,37], three *6-OST* genes [38] and five *3-OST* genes [39]. In continuing studies, these isoforms have been shown to have subtly different

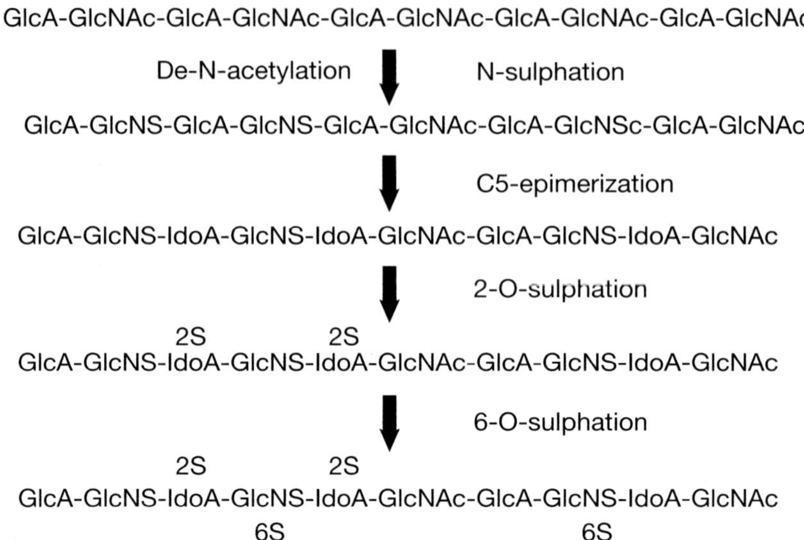

Figure 1 Some of the modification steps common in HS chain biosynthesis. The region shown would be a relatively highly sulphated section. Each modification step only affects a subset of possible substrates. Not all possible modifications are shown. For example, occasional GlcA residues may also become 2-O-sulphated, and GlcNAc residues may become 6-O-sulphated, but these events are not common in most HS chains. For this reason, 3-O-sulphation of GlcNS residues is also omitted. See text for more details.

substrate specificities (they recognize contrasting sequences and sulphation patterns within the nascent chain [38]), or to have alternative catalytic activities [37]. Significantly, these variants have been shown to have a clear, tissue-specific expression pattern, with some isoforms expressed much more widely than others. This provides a very attractive explanation for the observed variation in HS structure between tissues.

HS-pols

A variety of experimental systems has been used to generate animals mutant for components of the HS biosynthetic pathway. Most of these have been reviewed elsewhere, and owing to space restrictions, will not be discussed here [7,11,12,19,20]. With regard to mouse mutants, the most severe phenotype observed to date is that of the *EXT-1* (one of the HS-pol isoforms) knockout [40]. Heterozygotes reach adulthood with no observable defects, although they were shown to have decreased levels of HS (<50% of the wild-type level). Homozygote mutants, in contrast, were unable to complete gastrulation, and showed delayed and abnormal development of both the visceral endoderm and mesoderm [35]. These results highlight the absolute requirement for HS at a very early stage in development. They also indicate that, although there is a

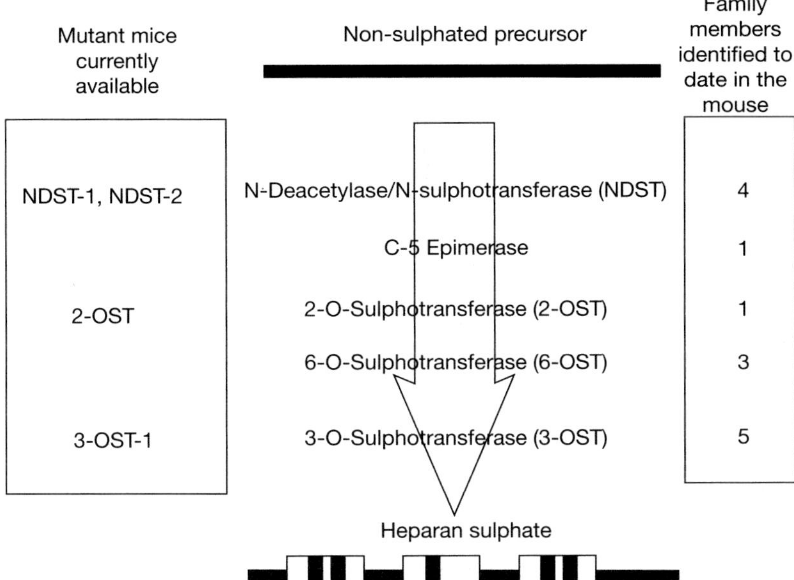

Figure 2 Enzymes involved in HS biosynthesis. Many of the HS biosynthetic enzymes exist as multiple gene families. The only exceptions to this are the epimerase and the 2-OST. For some of these enzymes, enzyme-deficient mice have been generated, as indicated in the panel on the left.

potential for redundancy in the HS biosynthetic process as, for example, both EXT1 and EXT2 are able to act as HS-pol enzymes, this is not realized. An explanation for this was suggested by McCormick et al. [41] who showed that a hetero-oligomeric complex of both EXT1 and EXT2 was required to allow functional localization of the complex to the Golgi apparatus. This is an area of particular interest, as defects in the human homologues of these genes are associated with the development of benign bony tumours (hereditary multiple exostoses) and an increased risk of chondrosarcomas and osteosarcomas [42]. The role of EXT1 and EXT2 remains controversial; however, as discussed in [43], new information from hereditary multiple exostoses patients, *EXT* knockout animals and cell lines lacking *EXT*s all need to be combined to understand fully the polymerization step in HS biosynthesis.

NDSTs

At present, knockout mice have been generated for two of the *NDST* isoforms, *NDST-1* and *NDST-2* [44–46]. Both of these isoforms have a very widespread expression pattern, which contrasts with *NDST-3* or *NDST-4*, whose expression is significantly more restricted [37]. Given this expression pattern, the phenotype of the *NDST-2* knockout mice is surprisingly cell-specific, with defects limited primarily to connective tissue-type mast cells [45,46].

In the *NDST-2* mutants of these cells, the dense granules that characterize the normal phenotype are replaced with a large number of empty vacuoles, with only a few granules remaining. In addition, the levels of mast cell proteases (chymases and tryptases) were significantly reduced, as were histamine levels; this loss was associated with a complete lack of heparin in these cells. When other tissues known to have high levels of NDST-2 mRNA were analysed, neither the liver nor the kidneys showed any abnormal HS structure. NDST-2 had, at an early stage in the investigation of HS biosynthesis, been suggested as the significant discriminator between HS and heparin biosynthesis [47]. The NDSTs are assumed to be the key determinants in laying down the N-sulphation pattern upon which all other modifications depend, and it is this pattern that is the principal characteristic used to categorize HS as a separate entity from heparin [3]. However, this view of the activity of NDSTs was revised after the observation of the widespread expression of NDST-2 mRNA, including tissues that lack heparin [4,48]. Mast cells are unusual in that they show high levels of NDST-2 mRNA with no detectable *NDST-1* [4]. It is possible therefore that NDST-2 alone is responsible for the biosynthesis of heparin in mast cells (it is not known if either *NDST-3* or *NDST-4* are expressed in mast cells). In other tissues, the potential contribution of NDST-2 to HS biosynthesis is either easily compensated for by the other NDSTs, or its function is redundant. Interestingly, the loss of heparin in the *NDST-2* knockout mice was not associated with a pro-coagulant phenotype, confirming that, *in vivo*, HS fulfils the required anti-coagulant function mimicked by pharmaceutical heparin [49,50].

As might be expected given the discussion above, the *NDST-1* knockout mice demonstrated a much more dramatic alteration in HS structure with most tissues having significantly decreased levels of N-sulphation [44]. Despite this, the phenotype of these mice was rather restricted, with mice dying soon after birth owing to a lung abnormality in which there is an increased number of type II endothelial cells in the lung, and decreased secretion of surfactant. Other non-penetrant defects were also apparent. It appears therefore that unlike heparin, the biosynthesis of HS (or at least a low-sulphated form of HS) can be, to some extent, catalysed by more than one NDST. The results also suggest that the abnormal HS retains many of the characteristics of 'normal' HS in allowing the development of the mice to proceed until birth, unlike animals lack that HS altogether.

The NDST mutants illustrate how knockout mice have been used to further investigate a system that had been analysed previously in great detail *in vitro*. In most cases, the results were not expected, and are a significant step forward in trying to understand how multiple enzymes, all of which are capable of catalysing the same event, interact to generate two very different polysaccharides.

3-OSTs

Within the extensive 3-OST family, 3-OST1 occupies a unique position as a resident of the Golgi lumen, as it lacks a transmembrane domain [51]. 3-OST1 is also the only 3-OST to participate in the formation of the antithrombin-binding region of HS [52]. It was not surprising therefore that *Hs3st1*$^{-/-}$ mice produced much less antithrombin-binding HS (75–98% less) [53]. The unexpected result of this study was that this did not lead to a pro-coagulant phenotype in the mice. Other, unanticipated, phenotypes were observed, but none of these were related to coagulation.

2-OSTs

Unlike the NDSTs, the 6-OSTs and the 3-OSTs, two HS-modifying enzymes have so far been identified as single isoforms; the epimerase [54] and the 2-OST (encoded by *Hs2st*) [55]. Interestingly, although IdoA and 2-O-sulphation have been found in all naturally occurring HS analysed to date, the amounts and patterns of these modifications vary widely between HS from different sources [23]. It seems unlikely that these differences are the result of differential expression of enzyme isoforms, so the patterns formed by the epimerase and the 2-OST must be controlled in a different manner.

In an alternative strategy to the conventional targeted knockout techniques used to generate the *NDST-1*$^{-/-}$ and *NDST-2*$^{-/-}$ mice described above, *Hs2st* mutant mice were generated by gene-trap mutagenic screening, which was used to identify genes that are important in early embryonic development [56]. The insertion of the gene-trap vector mutates, tags for cloning and reports on the expression pattern of genes. This latter feature was used to demonstrate the very widespread expression pattern of *Hs2st* in many stages of the developing mouse, with *Hs2st* absent in only very selected spatio-temporal locations such as the developing heart and the late-stage kidney [57].

Mice with only one functional copy of *Hs2st* were normal and fertile [56]; however, homozygous mutants died at birth, suffering bilateral renal agenesis (in >95% of pups). Other defects were also apparent, including skeletal, ocular and neural abnormalities. This complex, but relatively restricted, phenotype is particularly intriguing as 2-O-sulphation has been identified as a key component of many of the recognition sites within HS for binding to various ligands [58–60]. If, as would be expected, all 2-O-sulphation is lost in the mutants, a large proportion of the HS-dependent interactions should fail, and the effects would be predicted to be severe, being more similar to the *EXT-1* knockout, without the modulating effect of other isoforms, as seen in the *NDST-1*-deficient mice.

We have analysed HS from the *Hs2st* mutant mice, and compared it with that from the wild-type mouse [57]. The incorporation of a reporter gene into the gene-trap vector was particularly useful as it allowed 12.5-day embryos to be genotyped quickly and pooled accordingly. From these pools, embryonic fibroblasts were prepared and maintained in culture for 1–2 weeks. Metabolically radiolabelled HS was extracted from the cells and subjected to detailed compositional analysis. This analysis showed that the incorporation of the gene-trap vector had indeed created a null mutation, with no 2-O-sulphation observed in the HS from the *Hs2st*$^{-/-}$ cells. However, 6-O-sulphation and N-sulphation were both increased in this abnormal HS. These increases were made more dramatic because they were almost exclusively restricted to the S-domains of the chain. In the HS from *Hs2st*$^{-/-}$ cells, there was an increase in 6-O-sulphation of HexA-GlcNS, but not HexA-GlcNAc. Similarly, although the increase in N-sulphation was relatively modest (from 44% to 49% of all disaccharides) this was all contained within the S-domains that are composed of HexA-GlcNS-HexA-GlcNS repeat regions. This increase was associated with a corresponding decrease in the occurrence of mixed sequences of GlcNAc-GlcA-GlcNS-HexA-GlcNAc type. The enhanced N- and 6-O-sulphation was therefore restricted to the very regions from which 2-O-sulphation is predominantly lost (Figure 3). When the domain patterns of the HS chains from the mutant and wild-type are compared, it is apparent that the increased 6-O- and N-sulphation in the mutant has the effect of compensating, at least in terms of charge, for the loss of 2-O-sulphation. These results are in good agreement with a previously published study of HS synthesized by a CHO cell line that lacks 2-OST function, which was isolated from a chemical mutagenesis screen [61]. The discrete S-domain structure is one of the most conserved and characteristic features of HS, and the one that most dramatically sets it apart from heparin [3]. It is interesting therefore that this pattern is maintained (in a slightly expanded form) even after removing 2-O-sulphation, one of its key components. It was not possible to directly study the IdoA/GlcA ratios of the wild-type and mutant HS in the *Hs2st*$^{-/-}$ study, but this has been examined in mutant CHO cells, in which no difference in the degree of epimerization was observed [61].

The increased N- and 6-O sulphation in HS from either the *Hs2st*$^{-/-}$ mutant or from the *Hs2st*-deficient CHO cells is difficult to explain fully using the biosynthetic plan outlined in Figures 1 and 2. 2-O-sulphation occurs midway through the modification pathway and predominantly on IdoA residues generated by the epimerase, which in turn requires the action of an NDST to form its substrate. It seems logical that an early event will affect a later event, and therefore it is possible that the increased 6-O-sulphation in the HS from

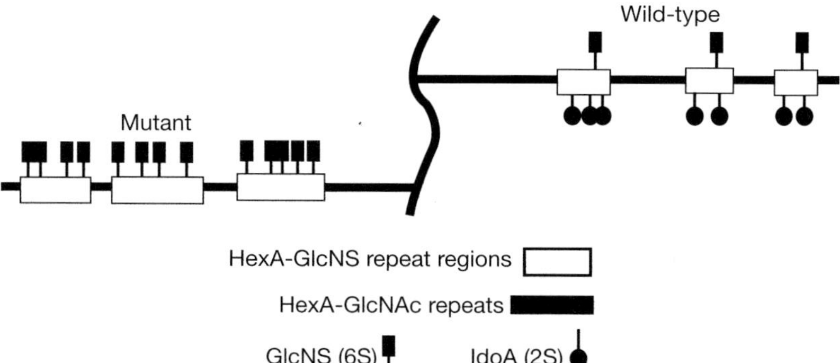

Figure 3 Retention of domain organization in HS from *Hs2st*$^{-/-}$ cells. HS from the mutant mice was found to contain no 2-O-sulphation, but increased 6-O- and N-sulphation. Most of this increase occurred within the S-domains, compensating, at least in terms of charge, for the loss of 2-O-sulphation.

Hs2st$^{-/-}$ cells is the result of a 2-*O*-sulphate-deficient intermediate forming a better substrate for whichever 6-OST catalyses the addition to HexA-GlcNS disaccharides. Using reverse transcriptase–PCR, we found that 6-OST1 (but not 6-OST2 or 6-OST3) was expressed in both CHOK1 (wild-type) and pgsF-17 (*Hs2st* mutant) cell lines (C.L.R. Merry, unpublished work). In a similar study using the mouse fibroblasts, all three isoforms were present in both wild-type and mutant cells, although 6-OST3 was only present at relatively low levels (C.L.R. Merry, unpublished work). This might suggest that 6-OST1 is responsible for the increased 6-O-sulphation in the HS from mutant cells, although the contribution of the other isoforms cannot be ruled out. This is in agreement with the published substrate specificity of 6-OST1, which shows a preference for IdoA-GlcNS over GlcA-GlcNS disaccharides [38].

It is more difficult to explain how a loss of 2-O-sulphation is associated with an increase in N-sulphation. It is possible that 2-O-sulphation is able to restrict N-sulphation in some way. In support of this idea, Aikawa et al. suggested that some of the NDST isoforms (NDST-3 and NDST-4) are particularly suited to acting on areas of the chain that have been modified previously [37]. The loss of 2-O-sulphation may therefore create more suitable substrates for these isoforms. An alternative explanation is suggested by recent studies, which have shown that various components of the HS biosynthetic machinery appear to require co-expression for full functionality. This has been shown for the polymerases (EXT1 and EXT2, as discussed above and in [41]), the epimerase and the 2-OST [62]. Although NDST-1 was not found to form part of the complex in these experiments, this may not be the case for other NDST isoforms. It is possible therefore that instead of a step-wise progression through discrete enzymic processes, with the substrate specificities of the later enzymes determining their sites of action, other factors such as position and substrate competition may need to be taken into consideration. If various combinations of enzymes do form hetero-oligomeric complexes, these are likely to be severely disrupted in the knockout mutants, with the potential for altering the activity of enzymes that act 'earlier' as well as 'later' in the biosynthetic pathway. It is difficult to predict what effect positional reorganization of the enzymes might have on the HS made by a cell, but this may be a significant factor in determining HS structure. Future studies using cells and animals lacking components of the HS biosynthetic pathway will, therefore, be significantly enhanced by a thorough structural analysis of the HS, detailing patterns of modification as well as composition. Using this approach, it should be possible to build a clearer picture of how the various enzymes interact and this, in turn, will hopefully lead to a better understanding of the controlled complexity of HS structure.

References

1. Lindahl, U., Kusche-Gullberg, M. and Kjellén, L. (1998) J. Biol. Chem. **273**, 24979–24982
2. Esko, J. and Lindahl, U. (2001) J. Clin. Invest. **108**, 169–173
3. Gallagher, J.T. and Walker, A. (1985) Biochem. J. **230**, 665–674
4. Kusche-Gullberg, M., Eriksson, I., Pikas, D.S. and Kjellén, L. (1998) J. Biol. Chem. **273**, 11902–11907

5. Stringer, S.E., Mayer-Proschel, M., Kalyan, A., Rao, M. and Gallagher, J.T. (1999) J. Biol. Chem. **274**, 25455–25460
6. Lander, A. (1998) Matrix Biol. **17**, 465–472
7. Kjellén, L. and Lindahl, U. (1991) Annu. Rev. Biochem. **60**, 443–475
8. Filmus, J. and Selleck, S.B. (2001) J. Clin. Invest. **108**, 497–501
9. Bernfield, M., Kokenyesi, R., Kato, M., Hinkes, M.T., Spring, J., Gallo, R.L. and Lose, E.J. (1992) Annu. Rev. Cell Biol. **8**, 365–393
10. Iozzo, R.V. (1998) Annu. Rev. Biochem. **67**, 609–652
11. Selleck, S.B. (2000) Trends Genet. **16**, 206–212
12. Perrimon, N. and Bernfield, M. (2000) Nature (London) **404**, 725–728
13. Gallagher, J.T. (1994) Eur. J. Clin. Chem. Clin. Biochem. **32**, 239–247
14. The, I., Bellaiche, Y. and Perrimon, N. (1999) Mol. Cell **4**, 633–639
15. Saksela, O., Moscatelli, D., Sommer, A. and Rifkin, D.B. (1988) J. Cell Biol. **107**, 743–751
16. Lyon, M., Deakin, J.A. and Gallagher, J.T. (1994) J. Biol. Chem. **269**, 11208–11215
17. Forsberg, E. and Kjellén, L. (2001) J. Clin. Invest. **108**, 175–180
18. Selleck, S. (1998) Matrix Biol. **17**, 473–476
19. Esko, J.D. (1991) Curr. Opin. Cell Biol. **3**, 805–816
20. Selleck, S. (2001) Semin. Cell Dev. Biol. **12**, 127–134
21. Sanderson, R.D., Turnbull, J.E., Gallagher, J.T. and Lander, A.D. (1994) J. Biol. Chem. **269**, 13100–13106
22. Maccarana, M., Sakura, Y., Tawada, A., Yoshida, K. and Lindahl, U. (1996) J. Biol. Chem. **271**, 17804–17810
23. Safaiyan, F., Lindahl, U. and Salmivirta, M. (2000) Biochemistry **39**, 10823–10830
24. Arikawa-Hirasawa, E., Watanabe, H., Takami, H., Hassell, J.R. and Yamada, Y. (1999) Nat. Genet. **23**, 354–358
25. Ishiguro, K., Kadomatsu, K., Kojima, T., Muramatsu, H., Nakamura, E., Ito, M., Nagasaka, T., Kobayashi, H., Kusugami, K., Saito, H. and Muramatsu, T. (2000) Dev. Dyn. **219**, 539–544
26. Woods, A. and Couchman, J.R. (2001) Curr. Opin. Cell Biol. **13**, 578–583
27. Cano-Gauci, D.F., Song, H.H., Yang, H., McKerlie, C., Choo, B., Shi, W., Pullano, R., Piscione, T.D., Grisaru, S., Soon, S. et al. (1999) J. Cell Biol. **146**, 255–264
28. Silbert, J.E. (1963) J. Biol. Chem. **238**, 3542–3546
29. Esko, J.D. (1991) Curr. Opin. Cell Biol. **3**, 805–816
30. Merry, C.L., Lyon, M., Deakin, J.A., Hopwood, J.J. and Gallagher, J.T. (1999) J. Biol. Chem. **274**, 18455–18462
31. Tsuda, H., Yamada, S., Yamane, Y., Yoshida, K., Hopwood, J.J. and Sugahara, K. (1996) J. Biol. Chem. **271**, 10495–10502
32. Lind, T., Lindahl, U. and Lidholt, K. (1993) J. Biol. Chem. **268**, 20705–20708
33. Wei, Z., Swiedler, S.J., Ishihara, M., Orellana, A. and Hirschberg, C.B. (1993) Proc. Natl. Acad. Sci. U.S.A. **90**, 3885–3888
34. Orellana, A., Hirschberg, C.B., Wei, Z., Swiedler, S.J. and Ishihara, M. (1994) J. Biol. Chem. **269**, 2270–2276
35. Lin, X., Gan, L., Klein, W.H. and Wells, D. (1998) Biochem. Biophys. Res. Commun. **248**, 738–743
36. Stickens, D. and Evans, G.A. (1997) Biochem. Mol. Med. **61**, 16–21
37. Aikawa, J., Grobe, K., Tsujimoto, M. and Esko, J.D. (2001) J. Biol. Chem. **276**, 5876–5882
38. Habuchi, H., Tanaka, M., Habuchi, O., Yoshida, K., Suzuki, H., Ban, K. and Kimata, K. (2000) J. Biol. Chem. **275**, 2859–2868
39. Shworak, N.W., Liu, J., Petros, L.M., Zhang, L., Kobayashi, M., Copeland, N.G., Jenkins, N.A. and Rosenberg, R.D. (1999) J. Biol. Chem. **274**, 5170–5184
40. Lin, X., Wei, G., Shi, Z., Dryer, L., Esko, J.D., Wells, D.E. and Matzuk, M.M. (2000) Dev. Biol. **224**, 299–311

41. McCormick, C., Duncan, G., Goutsos, K.T. and Tufaro, F. (2000) Proc. Natl. Acad. Sci. U.S.A. **97**, 668–673
42. Duncan, G., McCormick, C. and Tufaro, F. (2001) J. Clin. Invest. **108**, 511–516
43. Senay, C., Lind, T., Muguruma, K., Tone, Y., Kitagawa, H., Sugahara, K., Lidholt, K., Lindahl, U. and Kusche-Gullberg, M. (2000) EMBO Rep. **1**, 282–286
44. Ringvall, M., Ledin, J., Holmborn, K., van Kuppevelt, T., Ellin, F., Eriksson, I., Olofsson, A.M., Kjellén, L. and Forsberg, E. (2000) J. Biol. Chem. **275**, 25926–25930
45. Forsberg, E., Pejler, G., Ringvall, M., Lunderius, C., Tomasini-Johansson, B., Kusche-Gullberg, M., Eriksson, I., Ledin, J., Hellman, L. and Kjellén, L. (1999) Nature (London) **400**, 773–776
46. Humphries, D.E., Wong, G.W., Friend, D.S., Gurish, M.F., Qiu, W.T., Huang, C., Sharpe, A.H. and Stevens, R.L. (1999) Nature (London) **400**, 769–772
47. Eriksson, I., Sandbäck, D., Ek, B., Lindahl, U. and Kjellén, L. (1994) J. Biol. Chem. **269**, 10438–10443
48. Toma, L., Berninsone, P. and Hirschberg, C.B. (1998) J. Biol. Chem. **273**, 22458–22465
49. Damus, P.S., Hicks, M. and Rosenberg, R.D. (1973) Nature (London) **246**, 355–357
50. Marcum, J.A. and Rosenberg, R.D. (1989) in Heparin: Chemical and Biological Properties; Clinical Applications (Lane, D.A. and Lindahl, U., eds), pp. 275–294, Edward Arnold, London
51. Shworak, N.W., Liu, J., Fritze, L.M., Schwartz, J.J., Zhang, L., Logeart, D. and Rosenberg, R.D. (1997) J. Biol. Chem. **272**, 28008–28019
52. Liu, J., Shworak, N.W., Fritze, L.M., Edelberg, J.M. and Rosenberg, R.D. (1996) J. Biol. Chem. **271**, 27072–27082
53. Shworak, N.W., Post, M., Enjoi, K., Christi, P., Lech, M., Beeler, D., Rayburn, H. and Rosenberg, R.D. (2000) Glycobiology **10**, Abstract 20
54. Li, J., Hagner-McWhirter, Å., Kjellén, L., Palgi, J., Jalkanen, M. and Lindahl, U. (1997) J. Biol. Chem. **272**, 28158–28163
55. Kobayashi, M., Habuchi, H., Yoneda, M., Habuchi, O. and Kimata, K. (1997) J. Biol. Chem. **272**, 13980–13985
56. Bullock, S.L., Fletcher, J.M., Beddington, R.S. and Wilson, V.A. (1998) Genes Dev. **12**, 1894–1906
57. Merry, C.L., Bullock, S.L., Swan, D.C., Backen, A.C., Lyon, M., Beddington, R.S., Wilson, V.A. and Gallagher, J.T. (2001) J. Biol. Chem. **276**, 35429–35434
58. Turnbull, J.E., Fernig, D.G., Ke, Y., Wilkinson, M.C. and Gallagher, J.T. (1992) J. Biol. Chem. **267**, 10337–10341
59. Kreuger, J., Salmivirta, M., Sturiale, L., Gimenez-Gallego, G. and Lindahl, U. (2001) J. Biol. Chem. **276**, 30744–30752
60. Conrad, H.E. (1998) Heparin-Binding Proteins, Academic Press, San Diego, CA
61. Bai, X. and Esko, J.D. (1996) J. Biol. Chem. **271**, 17711–17717
62. Pinhal, M.A., Smith, B., Olson, S., Aikawa, J., Kimata, K. and Esko, J.D. (2001) Proc. Natl. Acad. Sci. U.S.A. **98**, 12984–12989

Biochem. Soc. Symp. **69**, 59–72
(Printed in Great Britain)

5

Genomic analysis of C-type lectins

Kurt Drickamer[1] and Andrew J. Fadden

Glycobiology Institute, Department of Biochemistry, University of Oxford, South Parks Road, Oxford OX1 3QU, U.K.

Abstract

Many biological effects of complex carbohydrates are mediated by lectins that contain discrete carbohydrate-recognition domains. At least seven structurally distinct families of carbohydrate-recognition domains are found in lectins that are involved in intracellular trafficking, cell adhesion, cell–cell signalling, glycoprotein turnover and innate immunity. Genome-wide analysis of potential carbohydrate-binding domains is now possible. Two classes of intracellular lectins involved in glycoprotein trafficking are present in yeast, model invertebrates and vertebrates, and two other classes are present in vertebrates only. At the cell surface, calcium-dependent (C-type) lectins and galectins are found in model invertebrates and vertebrates, but not in yeast; immunoglobulin superfamily (I-type) lectins are only found in vertebrates. The evolutionary appearance of different classes of sugar-binding protein modules parallels a development towards more complex oligosaccharides that provide increased opportunities for specific recognition phenomena. An overall picture of the lectins present in humans can now be proposed. Based on our knowledge of the structures of several of the C-type carbohydrate-recognition domains, it is possible to suggest ligand-binding activity that may be associated with novel C-type lectin-like domains identified in a systematic screen of the human genome. Further analysis of the sequences of proteins containing these domains can be used as a basis for proposing potential biological functions.

Animal lectin structure and function

Complex glycans on cell surfaces, in the extracellular matrix and on soluble, secreted glycoproteins can carry information that must be decoded by animal lectins or sugar-binding proteins [1]. Recognition events between glycans and lectins mediate glycoprotein trafficking and turnover, as well as cell adhesion

[1]To whom correspondence should be addressed (e-mail kd@glycob.ox.ac.uk).

and cell–cell communication. Roles for animal lectins in cell–cell interactions in the immune system and in pathogen recognition in the innate immune response have been extensively investigated [2]. Similarly, the importance of sugars in the maturation and delivery of proteins in the luminal compartments of eukaryotic cells has been well established (see Chapter 6 in this volume) [3]. The dramatic effects on development caused by modifying or eliminating various types of glycosylation suggest that similar recognition events take place at various stages in development, although identification of the relevant lectins is not so far advanced [4]. Although it is tempting to speculate on the importance of glycan diversity in brain development, it has proven particularly difficult to identify receptors that might utilize this diversity in the process of brain development. Thus, although good examples of animal lectin function are available, the overall picture of what processes they may mediate remains unclear. The availability of the complete human genome sequence has created the opportunity to gain a broad perspective on carbohydrate-binding proteins.

A key finding in the animal lectin field has been the identification of discrete, modular domains that mediate carbohydrate recognition [5]. These domains fall into a number of different structural categories; the members of the largest and most complex group are designated C-type lectins because they require Ca^{2+} in order to bind glycan ligands. The carbohydrate-recognition domains (CRDs) within these proteins share a common overall fold, which was first described in the structure of an N-terminal fragment of rat serum mannose-binding protein [6]. As illustrated in Figure 1, the sugar binding takes place at a conserved Ca^{2+}-binding site and requires that key hydroxyl groups on the sugar ligands from a set of hydrogen bonds with acidic and amide amino-acid side chains on the protein. These amino-acid side chains and the sugar hydroxyl groups also form co-ordination bonds with the Ca^{2+}. The C-type CRDs represent a subset of a larger family of protein modules that are denoted C-type lectin-like domains (CTLDs) [5]. Many of the CTLDs bind proteins rather than sugar ligands, and this binding is often Ca^{2+}-independent (Figure 2).

Genomic screening for CTLDs

The conceptual approach to the identification of possible animal lectins is summarized in Figure 3. The first step is to identify protein modules that have structural features of C-type CRDs. Proteins containing CTLDs are examined further in two general ways. In the first, the CTLDs are separated from the remainder of the protein sequence and aligned with known CRDs. The presence of residues that form sugar-binding sites in *bona fide* CRDs is used to make predictions about possible sugar-binding activity. In the second, the sequences of the entire proteins are analysed to see the context of the putative CRD and to establish likely locations, usually in the plasma membrane, the extracellular matrix or as soluble secreted proteins. Other modules in the protein may also suggest possible biological functions of the molecule.

Some of the steps required to implement this general strategy are indicated in more detail in Figure 4. Our knowledge of CTLDs can be used to devise a systematic approach to the identification of genes encoding CTLDs.

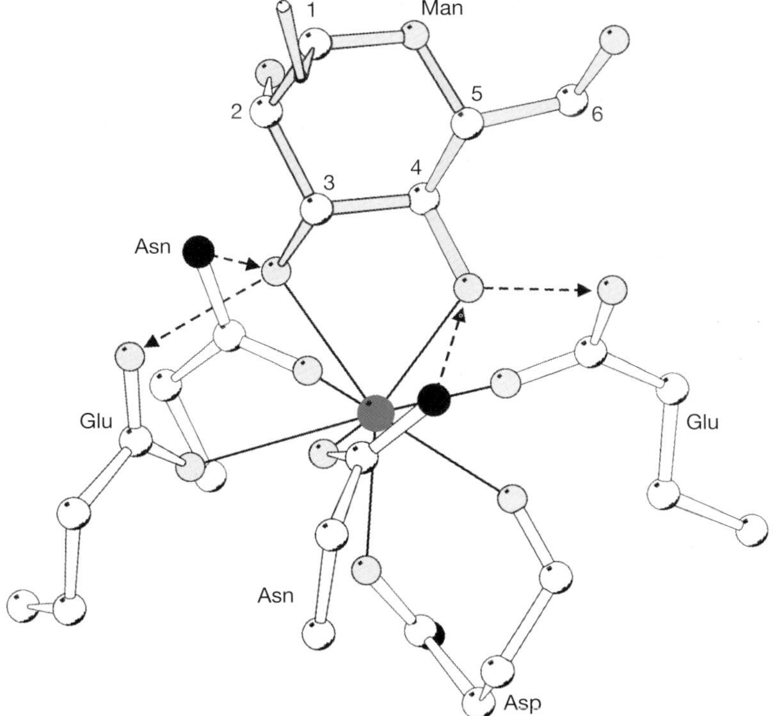

Figure 1 Ligand-binding site of the CRD of the prototypical C-type lectin, mannose-binding protein. Spheres represent carbon (white), oxygen (light grey), nitrogen (black) and Ca^{2+} (dark grey). Hydrogen bonds are shown as broken arrows and co-ordination bonds as solid lines.

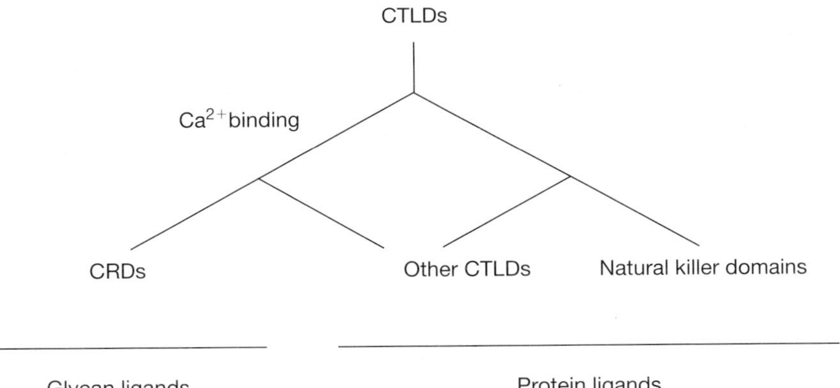

Figure 2 Relationship between CRDs and the larger family of CTLDs.

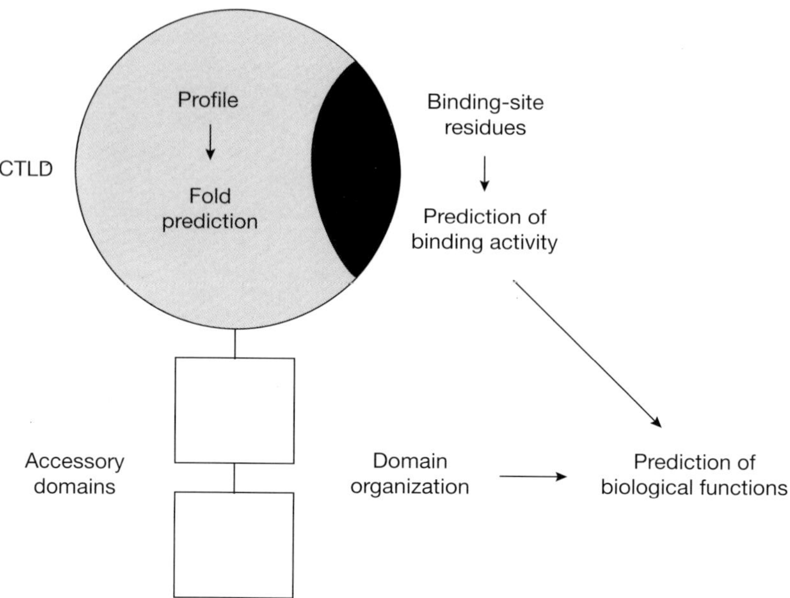

Figure 3 General strategy for identification of possible animal lectins and assignment of possible ligand-binding activities and biological functions.

The original motif used to spot domains of this type was based on a set of invariant and highly conserved residues that were characteristic of the members of the family that were known at the time. This approach has now been supplanted by more sophisticated computational methods involving domain profiles. Such profiles are generated from multiple sequence alignments of all known members of the domain family. Once the sequences are aligned optimally, the likelihood of each possible amino acid appearing at each position is tabulated to create the profile; for example, an invariant disulphide bond would be indicated by two positions at which the probability of cysteine residues occurring is 1.0 and the probability of each of the other 19 amino acids occurring is 0.

Predicted protein sequences are aligned with the profile to achieve the optimal match to the established frequencies of amino acids at each position. Based on alignments with randomized sequences, the statistical significance of the matching scores can be evaluated. The process can be iterated, with new sequences added to the profile, which is then refined to reflect the changed frequencies. Profiles that identify CTLDs have been established in several databases, including the PROSITE database set up at the Swiss Institute of Bioinformatics [7], the Pfam (protein families) database maintained at the Sanger Institute [8], and the InterPro and SMART databases that are run by the European Bioinformatics Institute [9,10].

The *Caenorhabditis elegans* and *Drosophila melanogaster* sequences have been completed to a very high degree of reliability, and annotation of possible genes is, for the most part, accurate [11,12]. The presence of introns is the major

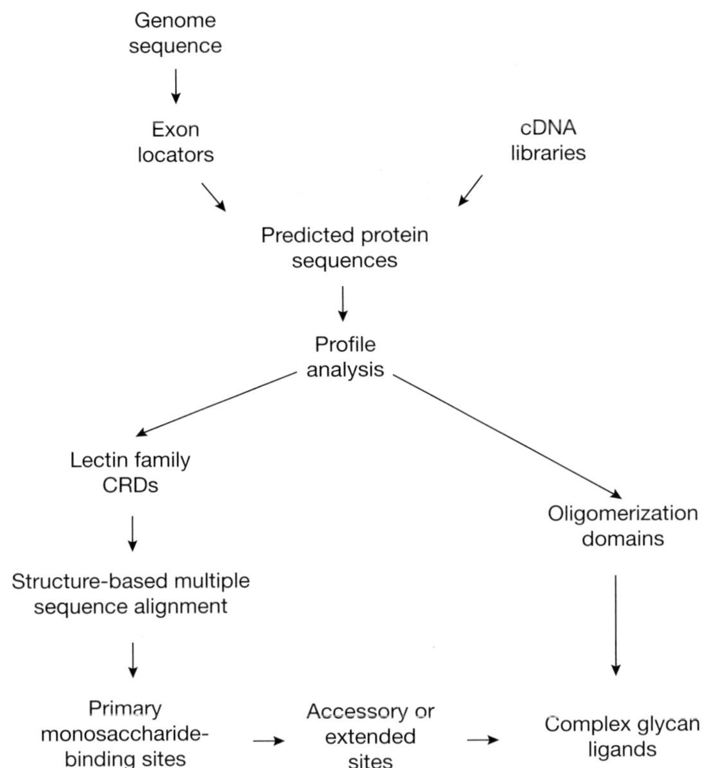

Figure 4 Details of screening procedure for analysing genome sequence information. The steps up to the profile analysis are performed in a mostly automated fashion by the curators of the sequence databases.

source of potential confusion since incorrectly predicted splicing events can lead to severe corruption of deduced protein sequences. The human genome sequence has yet to be established at the same degree of reliability. Thus, annotation in the Ensembl database at the European Bioinformatics Institute and the Sanger Institute is changing continuously as sequences are edited and corrected. Analysis of this sequence-in-progress has been supplemented by screening the database of cDNA sequences generated from various tissue libraries. No new proteins containing potential CTLDs came to light in the 3 months prior to the writing of this chapter, which suggests that coverage of the genome is nearly complete. Thus, it appears that there is a total of approximately 63 proteins containing CTLDs in the human genome.

Although the profiling methodology is quite sophisticated, sequences can be incorrectly identified as containing potential CTLDs. Examination of all of the sequences in the SwissProt database that are identified as matching one of the CTLD profiles suggests that 5–10% of the sequences that receive this annotation do not contain CTLDs. The absence of CTLDs can be demonstrated by repeating the profile scans, which in these cases fail to show even a statistically dubious match to the profiles. The reason for this discrepancy is not clear. It does seem that the mismatched sequences are often cysteine-rich and, in several cases, make good matches to other domain profiles. Alignments with these sequences confirm that they do not show the expected patterns of core hydrophobic residues that are required to form the CTLD fold. It is also possible that the profile screen can miss potential CTLDs; however, in terms of identifying potential sugar-binding proteins this is less of a problem. Highly divergent sequences are unlikely to form binding sites that are analogous to those in the known C-type lectins.

Identification of potential CRDs

Using our knowledge of how C-type CRDs bind to Ca^{2+} and glycans, it is possible to examine novel human CTLDs to see which ones are likely to bind these ligands (Figure 4). There are five amino-acid side chains that contribute ligands for the primary Ca^{2+}-binding site in most of the known C-type CRDs. These side chains (aspartate, asparagine, glutamate and glutamine) are polar and have the potential to form multiple hydrogen bonds. In a few cases, one of these ligands is missing and serine, threonine or a water molecule serves as a Ca^{2+} ligand. Predictions about the presence of this Ca^{2+} site can thus be made with some confidence.

Because of the way that sugar ligands interact with the conserved Ca^{2+} in the C-type CRDs, only those CTLDs that contain this site are considered as likely to display sugar-binding activity. In crystal structures of C-type CRDs complexed with saccharide ligands, the hydrogen-bonding scheme requires the presence of acid and amide side chains that form co-operative hydrogen bonds with the 3- and 4-hydroxyl groups of the sugar ligand, although the geometrical configurations of the bound sugars can differ. However, given that some of the known lectins have serine or threonine residues in at least some of these positions, it seems prudent to suggest that any CTLD with a probable conserved Ca^{2+}-binding site must be considered a potential sugar-binding protein. In cases in which the pattern of liganding side chains exactly matches a pattern associated with binding of a specific class of saccharide ligand, a reasonably strong prediction about probable selectivity of binding can be made.

The full selectivity and affinity of glycan binding by the C-type CRDs is achieved by extension and combination of the primary monosaccharide-binding sites [13–15]. Although a number of types of extended binding sites are known, it is difficult to make firm predictions from the relatively limited database. Similarly, oligomerization can often be predicted from the nature of accessory domains adjacent to the CTLDs, but the implications of such oligomers for ligand-binding activity are difficult to discern; for example, helical coiled-coil domains cause the association of three CRDs in both mannose-binding protein and the asialoglycoprotein receptor, yet the former binds clusters of widely spaced monosaccharides on cell surfaces and the latter binds closely spaced terminal sugars in a single N-linked glycan [16,17].

Groups of human proteins containing CTLDs

CTLD-containing proteins can be classified by defining the overall domain organization of the proteins in which they are found. The profile-screening approach is used to identify what other types of domains are found in addition to the CTLDs in these proteins. The proteins can also be analysed using simpler sequence motifs to identify hydrophobic signal sequences and transmembrane domains, as well as repeated patterns that are characteristic of the coiled coils of α-helices and collagenous domains.

When these approaches are applied to the CTLD-containing proteins, a relatively small number of different protein organizations are revealed. The 14 groups of CTLD-containing proteins defined in this way are summarized in Figure 5. The genomic screen has revealed almost no novel domain organizations, as all the groups except XIII and XIV were known previously. The numbers of proteins that fall into each group are indicated in Table 1.

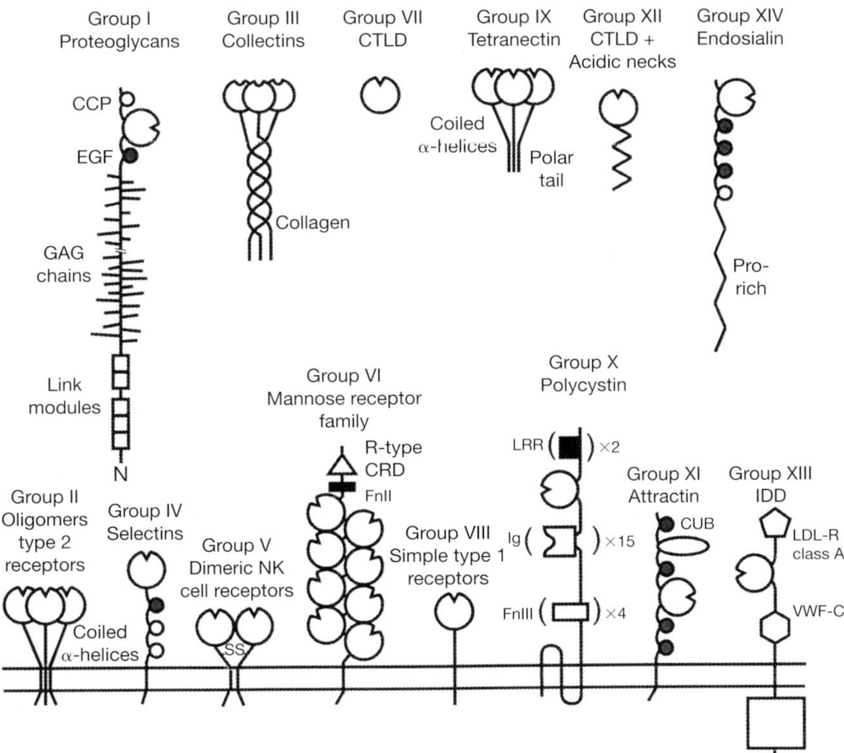

Figure 5 Domain organization of proteins that contain CTLDs. CCP, complement control protein; CUB, complement components C1r/C1s, Uegf and bone morphogenetic protein-1; EGF, epidermal growth factor; Fn, fibronectin; LDL-R, low-density lipoprotein receptor; LLR, leucine-rich repeat; SS, disulphide bond; VWF-C, von Willebrand factor.

Table 1 Summary of CTLDs in the human genome

Group	Number of members
I (proteoglycans)	4
II (type 2 receptors)	11
III (collectins)	6
IV (selectins)	3
V (natural killer cell receptors)	19
VI (mannose receptor family)	4
VII (free CTLDs)	5
VIII (layilin)	2
IX (tetranectin)	3
X (polycystin)	1
XI (attractin)	2
XII (myelin basic protein)	3
XIII (IDD)	1
XIV (endosialin)	1

Evolution of human CTLD sequences

One approach for the analysis of CTLD evolution is to analyse the degree of sequence similarity between CTLDs. In order to make good comparisons, it is necessary to identify the CTLD-coding sequences precisely. The end points of CRDs have been well defined using protease-resistance and expression studies [18]. The loop-out topology of the C-type CRDs brings the N- and C-terminal ends of the domain close to each other as a pair of β-strands that end in register at the domain boundaries [6]. The limits of the CRD also correspond to exon boundaries in many of the genes that encode these proteins [19]. Thus, it is possible to excise the regions of protein sequences that correspond to CTLDs and then to use multiple sequence alignment and cluster analysis to make comparisons that reflect only the evolution of the CTLDs and not their assortment with other protein modules. Dendrograms describing the relative similarities of the different domains reveal that many of the CTLDs fall into well-defined groups of relatively closely related domains. Remarkably, CTLDs in each of these groups always derive from proteins in the same structural groups defined by the modular organization of the parent proteins [19]. Thus, for the most part, it can be concluded that there was a single precursor for members of each group. A few CTLDs are too divergent to allow a robust demonstration that they are particularly closely related to any other members of the family. Most of the proteins containing such divergent CTLDs have unique domain organization.

Of the 63 proteins that contain CTLDs, approximately half are likely to function as lectins. The sugar-binding domains can be classified into two broad groups, based on the relative orientations of the 3- and 4-hydroxyl groups in the monosaccharides that interact with the primary ligand-binding site. Approximately two-thirds of the lectins are expected to bind mannose (Man) and *N*-acetylglucosamine, while the others bind galactose (Gal) and *N*-acetylgalactosamine; these two broad-specificity classes are referred to as Man-type and Gal-type respectively. In some cases, all members of a structural group show similar ligand-binding characteristics; for example, all of the collectins (Group III) bind Man-type sugars. In several cases, however, some members of a group bind sugars while other members of the group do not. Of particular interest are the members of Group II, which are clearly closely related to each other in terms of sequence, although some bind Man-type sugars while others bind Gal-type sugars.

Comparisons with model organisms

Because many developmental processes are much easier to analyse in relatively simple model organisms, the study of sugar-recognition molecules in these organisms could contribute significantly to our understanding of how glycans serve as recognition markers during development. For this reason, extensive studies of the CTLDs in *C. elegans* and *Drosophila* have been undertaken [20,21]. Unfortunately, the results suggest that CTLDs in these invertebrates have evolved along rather different lines compared with the mammals.

The differences between the CTLDs in different parts of the animal kingdom can be seen in two ways. Sequence comparisons between the CTLDs reveal that all of the *Drosophila* CTLDs are more related to each other than they are to any of the *C. elegans* or human CTLDs. Similarly, the *C. elegans* CTLDs form an evolutionary cluster that is separate from those in the other organisms. Comparison of the domain architecture of proteins containing CTLDs reveals that the only type of domain organization common to both mammals and either of the invertebrates is that seen in Group VII: CTLDs in isolation with no accessory domains. The organization of the CTLD-containing proteins in *C. elegans* is completely different to that in *Drosophila*. Therefore, it is not possible to identify clear orthologues of any of the human CTLD-containing proteins in the model organisms.

These results indicate that CTLDs have radiated independently in the lineages leading to humans, fruit flies and nematode worms. It is therefore interesting that there are some CTLDs in each of the invertebrate model organisms that are known lectins or are predicted from sequence analysis to have sugar-binding sites similar to those in human C-type CRDs. This finding indicates that sugar-binding activity must have arisen independently in each of the lineages of CTLDs, as indicated in Figure 6. This 'convergent' evolution of a sugar-binding site in the CTLD framework on multiple occasions probably reflects its relative simplicity, since only a few critical amino acids need to be present to form a monosaccharide-binding site. Another striking feature of the CTLDs in *C. elegans* and *Drosophila* is the relatively large number of these domains relative to genome size (Table 2). Moreover, the number of CTLDs that is likely to bind sugar is not correlated with the overall developmental complexity of the organisms, as the number of CTLDs predicted to bind sugars is substantially larger in *C. elegans* than in *Drosophila*.

Predictions about sugar-binding activity are, necessarily, rather tentative, so the likely presence of the conserved Ca^{2+}-binding site warrants investigation of the possible sugar-binding activity. The small pool of potential sugar-bind-

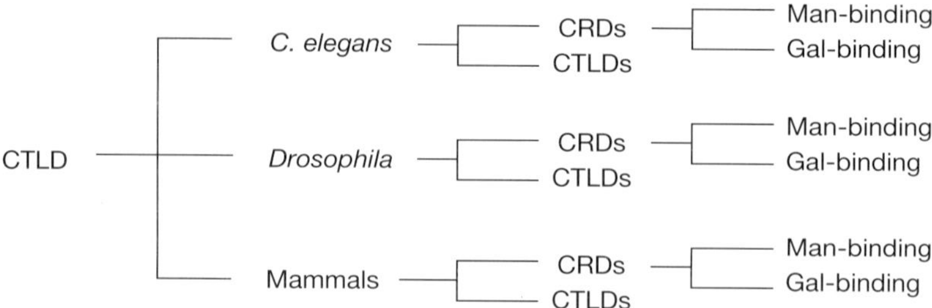

Figure 6 Evolution of sugar-binding activity in multiple lineages of CTLDs.

ing CTLDs in *Drosophila* makes it an attractive target for a functional genomics pilot project, in which the CTLDs are systematically expressed and their sugar-binding activity is analysed. A strategy for this type of study is outlined in Figure 7. cDNAs libraries from various developmental stages can be used as templates for the amplification of the CTLD-encoding regions, which can then be expressed in a bacterial system in which most CTLDs have been shown to fold directly after secretion into the periplasm. The activity of the proteins expressed in this way can be examined initially by affinity chromatography on immobilized sugars, which also serves as a convenient way to purify correctly folded proteins that can be used in further tests of ligand-binding activity. Such studies will ultimately be undertaken on the set of human CTLDs as well.

Table 2 Comparison of lectins in various genomes.
R-type, ricin-type lectin, I-type, immunoglobulin superfamily lectin.

Genes	*C. elegans*	*Drosophila*	Human
Number of genes in each organism	19000	13600	≈35000
Calnexin	2	2	3
L-type	2	2	2
P-type	0	0	2
M-type	3	2	3
R-type			
Group I	11	13	12
Group II	0	0	4
Galectins			
Group I	2	0	7
Group II	5	3	4
Group III	2	0	1
Group IV	7	1	0
C-type	125	32	66
CTLDs	183	33	96
CRDs	19	6	32
I-type	0	0	10
Total lectin genes	53	29	76

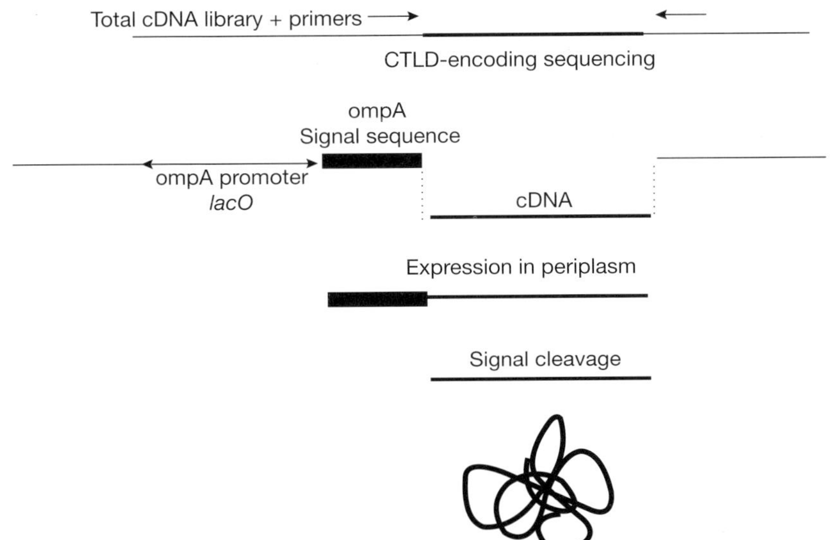

Figure 7 Strategy for demonstrating sugar-binding activity in CTLDs. *lacO*, operator sequence of the *Escherichia coli* lactose operon; ompA, outer membrane protein A.

Comparison with other types of animal lectins

Genome-wide studies of C-type lectins should be put in the overall context of genomic analysis of other structural categories of animal lectins [21]. The results of surveys of several genomes are summarized in Table 2. Different patterns of evolution of the intracellular and extracellular lectins are observed. The intracellular lectins involved in trafficking of proteins in between luminal compartments fall into the calnexin, L-type lectin and mannose 6-phosphate receptor families. Each of these families is limited to two or three members in each of the genomes examined.

In contrast with the intracellular lectins, the extracellular galectin and siglec families, like the C-type lectins, have expanded extensively within the vertebrate lineage. Among these families, the immunoglobulin-like siglecs (I-type lectins) are the least divergent in terms of sugar-binding activity, as they all bind glycans that contain sialic acid, and in overall organization, since they all consist of multiple immunoglobulin-type domains (see Chapter 7 in this volume). Siglecs have been identified only in vertebrates. The vertebrate galectins fall into three general structural categories, all of which are also found in *C. elegans*; only one of these groups is represented in *Drosophila*. A fourth group, not found in vertebrates, is found in *C. elegans* and *Drosophila*; for the most part, these proteins bind β-galactosides, although there are more divergent examples that bind other types of sugars including mannose. The radiation of extracellular lectins within the vertebrate lineage suggests that these proteins have diverged to fill novel carbohydrate-recognition roles in vertebrates.

One of the most striking features of the known families of lectins is that sugar-binding activity has evolved in the context of many different protein folds. The shallow binding sites and the weak affinities for monosaccharides that characterize these binding sites have apparently evolved on multiple occasions. This observation suggests that there may be additional protein folds that can accommodate sugar-binding activity in human proteins; two examples of such folds have emerged relatively recently.

R-type CRDs are related to the sugar-binding domains of ricin. These CRDs are unusual in that they are found in bacteria as well as in plants and animals. In all cases, they have a targeting function. In ricin, the R-type CRD targets transport of a ribosome-inactivating toxin for transport into the cytoplasm via a retrograde pathway through endosomes, the Golgi apparatus and the endoplasmic reticulum [22]. The R-type CRDs in bacterial proteins direct hydrolytic enzymes to their polysaccharide targets, e.g. cellulose [23]. In humans, R-type CRDs are found in two distinct types of proteins. The larger Group I consists of N-acetylgalactosaminyl transferases that initiate O-linked glycan synthesis [24]. The lectin domains appear to recognize previously glycosylated sites and allow the processive addition of glycans to nearby serine and threonine residues. The other proteins containing modules that resemble R-type CRDs are the members of the mannose receptor family (Group II). In addition to binding mannose and related sugars through C-type CRDs, the mannose receptor binds sulphated *N*-acetylgalactosamine residues via the R-type CRD [25]. Structural analysis suggests that the homologous domains from other members of the family almost certainly do not have this activity.

More recently, an additional family of intracellular lectins involved in glycoprotein trafficking has been identified. The term M-type lectin has been suggested for these proteins, which resemble mannosidases found in the endoplasmic reticulum, but lack key catalytic residues. These proteins function in the process of endoplasmic-reticulum-associated degradation, in which glycoproteins that have remained incorrectly folded in the endoplasmic reticulum for an extended period are directed to the cytoplasm for proteolytic degradation [26–29]. Like the other intracellular lectin families, the M-type lectin family is modest in size and has not been extensively elaborated in vertebrates compared with invertebrates.

The relatively recent discovery of R- and M-type lectins suggests that families of proteins that bind sugars will continue to be identified. Indeed, other families with such activities, e.g. the interleukins and several types of proteins that bind glycosaminoglycans, have already been described [30,31]. It is clear that sugar-binding activity can arise in the context of many different protein backgrounds, so that the profile-scanning approach would not be useful in detecting such proteins. In addition, sugar-binding proteins are not characterized by simple, contiguous peptide sequence motifs, which reflects the fact that sugar-binding sites arise from the interaction of residues that are close in space, but are not necessarily close in the primary structure of the protein. Thus, for the most part, detection of novel sugar-binding protein families will be an empirical matter.

World-wide web-based resource for animal lectins

The types of bioinformatics and functional genomics discussed in the preceding sections are directed towards an understanding of what genes encode glycan-binding proteins and what their potential glycan ligands may be. As indicated in Figure 3, the domain architecture of proteins containing potential sugar-binding domains may provide clues about their biological functions. However, one of the great advantages of the genomic approach to analysis of sugar-binding proteins is that it will facilitate application of genetic approaches to address issues of function.

An important genetic approach will be analysis of phenotypes associated with mutations in lectin genes. In many cases, identification of such mutations will follow from studies of disease conditions that can be mapped using positional cloning techniques. In order to make use of the information derived from such studies, it is important that the properties of candidate genes will be readily accessible to investigators carrying out mapping studies. For this reason, a database of sugar-binding proteins has been established as a genomic resource for animal lectin studies. The database is still undergoing development, but in its present version it is accessible on the world-wide web (http://ctld.glycob.ox.ac.uk).

There are several aspects to the current database. Part of the information is general background on animal lectin structures and known functions. A second section is devoted to compilation of the proteins containing CTLDs in humans and in model organisms. In its current form, the database is orientated toward structural analysis, but information on function at the level of glycan-binding activities and broader biological functions is also being incorporated.

In addition to providing a link between the work in glycobiology and the world of genomics, the lectin database is intended to assist in the co-ordination of work within the field of carbohydrate recognition. The relatively modest number of lectins that have been identified suggests that, although complex, the universe of protein–carbohydrate interactions is finite. As the identification phase of animal lectin biology comes to a close, the accumulated data on the lectins and their glycan ligands will facilitate the much larger task of understanding their biological functions.

This work was supported by the Wellcome Trust and the Biotechnology and Biological Sciences Research Council.

References

1. Drickamer, K. and Taylor, M.E. (1993) Annu. Rev. Cell Biol. **9**, 237–264
2. Weis, W.I., Taylor, M.E. and Drickamer, K. (1998) Immunol. Rev. **163**, 19–34
3. Ellgaard, L., Molinari, M. and Helenius, A. (1999) Science **286**, 1882–1888
4. Muramatsu, T. (2000) J. Biochem. (Tokyo) **127**, 171–176
5. Drickamer, K. (1999) Curr. Opin. Struct. Biol. **9**, 585–590
6. Weis, W.I., Drickamer, K. and Hendrickson, W.A. (1992) Nature (London) **360**, 127–134
7. Hofmann, K., Bucher, P., Falquet, L. and Bairoch, A. (1999) Nucleic Acids Res. **27**, 215–219

8. Bateman, A., Birney, E., Durbin, R., Eddy, S.R., Howe, K.L. and Sonnhamme, E.L. (2000) Nucleic Acids Res. **28**, 263–266
9. Apweiler, R., Attwood, T.K., Bairoch, A., Bateman, A., Birney, E., Biswas, M., Bucher, P., Cerutti, L., Corpet, F., Croning, M.D.R. et al. (2001) Nucleic Acids Res. **29**, 37–40
10. Schultz, J., Copley, R.R., Doerks, T., Ponting, C.P. and Bork, P. (2000) Nucleic Acids Res. **28**, 231–234
11. The *C. elegans* Sequencing Consortium (1998) Science **282**, 2012–2018
12. Adams, M.D., Celniker, S.E., Holt, R.A., Evans, C.A., Gocayne, J.D., Amanatides, P.G., Scherer, S.E., Li, P.W., Hoskins, R.A., Galle, R.F. et al. (2000) Science **287**, 2185–2195
13. Ng, K.K.S. and Weis, W.I. (1997) Biochemistry **36**, 979–988
14. Kolatkar, A.R., Leung, A.K., Isecke, R., Brossmer, R., Drickamer, K. and Weis, W.I. (1998) J. Biol. Chem. **273**, 19502–19508
15. Feinberg, H., Mitchell, D.A., Drickamer, K. and Weis, W.I. (2001) Science **294**, 2163–2166
16. Weis, W.I. and Drickamer, K. (1994) Structure **2**, 1227–1240
17. Lee, R.T., Lin, P. and Lee, Y.C. (1984) Biochemistry **23**, 4255–4261
18. Quesenberry, M.S. and Drickamer, K. (1991) Glycobiology **1**, 615–621
19. Drickamer, K. (1993) Prog. Nucleic Acid Res. Mol. Biol. **45**, 207–232
20. Drickamer, K. and Dodd, R.B. (1999) Glycobiology **9**, 1357–1369
21. Dodd, R.B. and Drickamer, K. (2000) Glycobiology **11**, 71R–79R
22. Day, P.J., Owens, S.R., Wesche, J., Olsnes, S., Roberts, L.M. and Lord, J.M. (2001) J. Biol. Chem. **276**, 7202–7208
23. Fujimoto, Z., Kuno, A., Kaneko, S., Yoshida, S., Kobayashi, H., Kusakabe, I. and Mizuno, H. (2000) J. Mol. Biol. **300**, 575–585
24. Hassan, H., Reis, C.A., Bennett, E.P., Mirgorodskaya, E., Roepstorff, P., Hollingsworth, M.A., Burchell, J., Taylor-Papadimitiou, J. and Clausen, H. (2000) J. Biol. Chem. **275**, 38197–38205
25. Fiete, D.J., Beranek, M.C. and Baenziger, J.U. (1998) Proc. Natl. Acad. Sci. U.S.A. **95**, 2089–2093
26. Jakob, C.A., Bodmer, D., Spirig, U., Battig, P., Marcil, A., Dignard, D., Bergeron, J.J.M., Thomas, D.Y. and Aebi, M. (2001) EMBO Rep. **2**, 423–430
27. Hosokawa, N., Wada, I., Hasegawa, K., Yorihuzi, T., Tremblay, L.O., Herscovics, A. and Nagata, K. (2001) EMBO Rep. **2**, 415–422
28. Nakatsukasa, K., Nishikawa, S., Hosokawa, N., Nagata, K. and Endo, T. (2001) J. Biol. Chem. **276**, 8635–8638
29. Braakman, I. (2001) EMBO Rep. **2**, 666–668
30. Cebo, C., Dambrouck, T., Maes, E., Laden, C., Strecker, G., Michalski, J.-C. and Zanetta, J.-P. (2001) J. Biol. Chem. **276**, 5685–5691
31. Mulloy, B. and Linhardt, R.J. (2001) Curr. Opin. Struct. Biol. **11**, 623–627

Biochem. Soc. Symp. **69**, 73–82
(Printed in Great Britain)

6

Lectins and protein traffic early in the secretory pathway

Hans-Peter Hauri[1], Oliver Nufer, Lionel Breuza, Houchaima Ben Tekaya and Lu Liang

Biozentrum, Klingelbergstrasse 70, University of Basel, Klingelbergstrasse 70, CH-4056 Basel, Switzerland

Abstract

Lectins of the early secretory pathway are involved in selective transport of newly synthesized glycoproteins from the endoplasmic reticulum (ER) to the ER–Golgi intermediate compartment (ERGIC). The most prominent cycling lectin is the mannose-binding type 1 membrane protein ERGIC-53 (ERGIC protein of 53 kDa), a marker for the ERGIC, which functions as a cargo receptor to facilitate export of an increasing number of glycoproteins with different characteristics from the ER. Two ERGIC-53-related proteins, VIP36 (vesicular integral membrane protein 36) and a novel ERGIC-53-like protein, ERGL, are also found in the early secretory pathway. ERGL may act as a regulator of ERGIC-53. Studies of ERGIC-53 continue to provide new insights into the organization and dynamics of the early secretory pathway. Analysis of the cycling of ERGIC-53 uncovered a complex interplay of trafficking signals and revealed novel cytoplasmic ER-export motifs that interact with COP-II coat proteins. These motifs are common to type I and polytopic membrane proteins including presenilin 1 and presenilin 2. The results support the notion that protein export from the ER is selective.

Lectins of the early secretory pathway

The early secretory pathway of animal cells is known to harbour four major lectins: calnexin, calreticulin, endoplasmic reticulum (ER)–Golgi intermediate compartment protein of 53 kDa (ERGIC-53) and VIP36 [1]. The related proteins calnexin and calreticulin bind monoglucosylated N-linked oligosacchrides and operate as chaperones in the quality control of newly synthesized glycoproteins in the ER (Figure 1) [2–5]. ERGIC-53 and its homologue VIP36 are mannose-

[1]To who correspondence should be addressed (e-mail hans-peter.hauri@unibas.ch).

specific lectins that cycle between membranes of the early secretory pathway [6,7]. They show sequence identity with the lectins of leguminous plant [8,9].

This chapter focuses on ERGIC-53, a long-living, non-glycosylated type I membrane protein with a large luminal domain and a 12-amino-acid cytoplasmic domain. ERGIC-53 forms disulphide-linked homodimers and homohexamers. It is expressed in all cell types and conserved in *Caenorhabditis elegans*, *Drosophila*, *Xenopus*, rat and humans. The non-essential yeast protein Emp47p is the yeast homologue of ERGIC-53, although it lacks some residues in the putative carbohydrate-recognition domain (CRD) that are critical for the binding of mannose [10]. The contribution of ERGIC-53 to the understanding of secretory pathway organization and of the mechanism of protein transport from the ER, as well as its relationship to VIP36 and a recently discovered ERGIC-53-like protein, ERGL [11], are discussed in this chapter.

ERGIC-53 and the organization of the early secretory pathway

ERGIC-53 can be visualized, by immunoelectron microscopy, primarily in characteristic tubulovesicular clusters near the Golgi apparatus, but also in the cell periphery [12–15]. Most clusters are localized near part-rough/part-smooth ER membranes known as transitional elements, which exhibit budding activity [15], and they accumulate newly synthesized secretory proteins in cells cultured at 15°C [16]. Based on these findings, the areas characterized by these ERGIC-53-positive clusters become known as the ERGIC [17]; thus, ERGIC-53 defines the ERGIC. Subdomains of ERGIC clusters also stain positive for COP-I proteins that form COP-I vesicles implicated in retrograde transport [15]. Because some areas of the clusters also stain positive for COP-II proteins, they are also termed ER-export complexes [14,18–20]. Recent analysis of ER to Golgi transport of vesicular stomatitis virus glycoprotein tagged with green fluorescent protein by video-enhanced fluorescence microscopy in living cells has led to the notion that the ERGIC elements are highly dynamic, short-lived transport intermediates that fuse to form the *cis*-Golgi [21–23]; however, studies with green fluorescent protein–ERGIC-53 in living cells provide a different picture. These studies revealed two types of labelled structures, rather long-lived larger elements and highly dynamic smaller, vesicular or tubular elements that are often found to depart from the large elements in random directions (H. Ben Tekaya and H.-P. Hauri, unpublished work). The larger elements correspond to the tubulovesicular ERGIC clusters and the dynamic elements to ERGIC-to-ER retrograde tubules seen by immunoelectron microscopy [15].

By integrating the studies on secretory proteins with those of the recycling protein ERGIC-53, we propose the following model of the organization of the early secretory pathway. Newly synthesized secretory proteins and cycling proteins, such as ERGIC-53, leave the ER at transitional elements and move to nearby ERGIC-clusters by a membrane budding process that is powered by COP-II coats. These clusters are the first way station at which anterograde (to the Golgi apparatus) and retrograde (to the ER) traffic separates. The majority of ERGIC-53 recycles from this point back to the ER by a

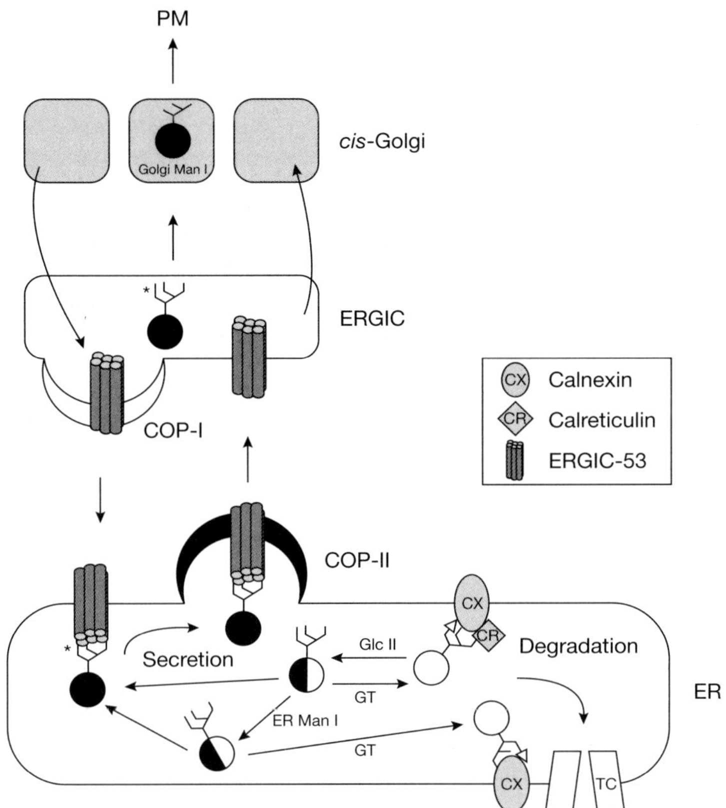

Figure 1 Recycling and cargo receptor function of ERGIC-53. After synthesis and trimming of the two outermost glucose residues of their N-glycans, many glycoproteins bind to the ER lectins calnexin and/or calreticulin, which recognize monoglucosylated N-glycans. Subsequently the glycoproteins are trimmed by glucosidase II (Glc II) and, if still incompletely folded (indicated in white), reglucosylated by UDP-glucose glycoprotein glucosyltransferase (GT), which redirects them to another cycle of quality control [2]. After prolonged residence time in the ER, ER mannosidase I (ER Man I) removes one mannose residue from the middle branch of the N-glycan. Incompletely folded, and thus reglucosylated, Man_8 glycoproteins are targeted for calnexin-dependent retrotranslocation through the translocation channel (TC) to the cytosol and subsequent degradation by the proteasome. By contrast, correctly folded proteins (marked in black) are no longer recognized by GT after deglucosylation by Glc II and are transport-competent. They may or may not undergo some additional trimming by mannosidase I and II before leaving the ER. Some of these Man_{7-9} glycan-bearing proteins (indicated by an asterisk) now bind to the lectin ERGIC-53, which recruits them to COP-II buds and thereby facilitates transport to the ERGIC. Dissociation of ERGIC-53 and its glycoprotein ligand occurs in the ERGIC and free ERGIC-53 recycles to the ER via COP-I vesicles. A minor fraction of ERGIC-53 escapes to *cis*-Golgi, but is retrieved. In the *cis*-Golgi, glycoproteins undergo further trimming to Man_5 prior to reglycosylation by Golgi glycosyltransferases and further transport along the secretory pathway. PM, plasma membrane.

COP-I-dependent process; cycling through the Golgi is only a minor route for ERGIC-53. In contrast with present thinking, the clusters are not consumed by this sorting process, but can accept several rounds of traffic from the ER and hence are not just transport containers. Anterograde traffic from the ERGIC involves highly mobile membranes that migrate to the Golgi, and retrograde traffic involves tubular membranes that ultimately fuse with the ER. We view the ERGIC as a scattered membrane compartment in the true sense, rather than a collection of transient transport containers consumed by one round of ER-to-Golgi traffic. This model can now be tested in living cells by simultaneous recording of anterograde and retrograde transported marker proteins. ERGIC-53 will be an invaluable tool in such experiments.

Functional dynamics of the lectin ERGIC-53

ERGIC-53 has several hallmarks of a transport receptor for glycoproteins. It binds to immobilized mannose via a CRD that has homology with lectins from leguminous plants, cycles between the ERGIC and ER, and possesses anterograde and retrograde transport signals that interact with COP-II and COP-I coats respectively. Indeed, non-functional ERGIC-53 leads to an impairment of glycoprotein transport from the ER. Interestingly, this defect is limited to a subgroup of glycoproteins including the lysosomal enzymes cathepsin C [24] and cathepsin Z [25], and the blood coagulation factors V and VIII [26]. Mutations in humans leading to non-functional ERGIC-53 cause a combined deficiency of factors V and VIII [27]. ERGIC-53 is also required for efficient transport of the brush-border membrane protein sucrase isomaltase (SI) (S. Mitrovic and H.-P. Hauri, unpublished work), and in cases of congenital SI deficiency with a block of misfolded SI in the Golgi, ERGIC-53 is also mislocalized to the Golgi [28], which is presumably due to prolonged interaction with its cargo. SI is the first membrane glycoprotein that has been found to require ERGIC-53 for efficient transport. The basis for selective cargo binding of ERGIC-53 remains to be elucidated.

In *C. elegans*, ERGIC-53 is expressed in all cells, with highest expression in those cells that are secreting actively. Functional inactivation of ERGIC-53 in *C. elegans* by the RNA interference approach leads to an egg maturation defect and reduced fertility (G. Cassata, F. Kuhn and H.-P. Hauri, unpublished work). This is in accord with the notion that ERGIC-53 is required for the secretion of selective proteins. *C. elegans* will provide an exciting system with which to dissect the function of ERGIC-53 by genetics.

Cross-linking studies in cell culture [25] suggest that ERGIC-53 operates as illustrated in Figure 1. ERGIC-53 binds to newly synthesized, correctly folded glycoproteins in the ER in a calcium-dependent process and recruits them to budding COP-II vesicles. Man_9 appears to be the preferred glycan structure for binding. Dissociation of cargo occurs in the ERGIC and does not require mannose trimming by Golgi mannosidases. Dissociation may be mediated by the ionic milieu in the ERGIC that is unknown. Free ERGIC-53 is subsequently recognized by COP-I and recycled back to the ER via COP-I vesicles for a next round of transport.

Trafficking signals of ERGIC-53 and ER export

To uncover signals that are required for protein cycling, we used an extended mutagenesis approach that revealed a complex interplay of ER retention, ER export and ER retrieval motifs. All domains of ERGIC-53 are required for correct targeting. The cytoplasmic domain comprises a C-terminal diphenylalanine ER-export motif and a dilysine ER retrieval signal [6,29]. The transmembrane domain contributes to both ER retention and ER export, and the luminal domain is required for oligomerization, which is another prerequisite for ER export (O. Nufer and H.-P. Hauri, unpublished work).

Interestingly, the diphenylalanine motif of ERGIC-53 is not strictly conserved during evolution [1], which suggests that other motifs may also function in ER export. A mutagenesis study of the two C-terminal phenylalanine residues has indeed confirmed this notion [30]. The diphenylalanine motif can be functionally substituted by a single phenylalanine or tyrosine residue at position 2 from the C-terminus, two leucine or isoleucine residues at positions 1 and 2, or a single valine residue at position 1. The valine motif is a transport signal in the true sense since it is transplantable to other reporter proteins, whereas the other motifs are not sufficient for mediating such a transport. However, all the transport motifs mediate COP-II binding in an *in vitro* assay with different preferences for COP-II subunits. Previous studies have reported a requirement of a C-terminal valine residue for efficient transport of transforming growth factor α, the membrane-type metalloproteinase 1 and the plasma membrane protein CD8 [31–33], which is in line with our conclusion that a C-terminal valine residue is a transport signal.

A search in non-redundant databases from SwissProt and trEMBL for human type I membrane proteins revealed that the newly found ER export motifs are common (Table 1); of the type I proteins, 19% carry such signals. Moreover, candidate amino acids appear more often in functional than in non-functional positions; for instance, a phenylalanine residue is found more frequently at position 2 than at position 1. Furthermore, a C-terminal valine residue at position 1 is found more often than would be expected theoretically. Similar results were obtained when analysing mammalian type I membrane proteins.

C-terminal ER-export motifs are also found in polytopic membrane proteins including presenilin 1 and presenilin 2. Various mutations in presenilins lead to early-onset Alzheimer's disease. Replacing the C-terminal amino acids of presenilin 1 or presenilin 2 with alanine results in significantly delayed ER-to-Golgi transport (Figure 2), which indicates that the C-terminal motifs that we have uncovered also contribute to efficient ER export of polytopic membrane proteins.

Our analysis of the diphenylanine motif of ERGIC-53 and its extension to other motifs increases considerably, what is at present, the small repertoire of ER-export motifs (Table 2) and strongly supports the notion of selective, signal-mediated protein export from the ER, a process that is mediated by signal–COP-II coat interaction.

Table 1 Frequency of motifs at C-terminus of human type I membrane proteins. Amino acid motifs at position 2 and position 1 from the C-terminus (>) are indicated. Amino acids in square brackets are accepted in this position. Amino acids in braces are not accepted in this position. Data taken from [30].

	Transport motif*	Number of hits (%)†
Functional transport motifs	{FY}V>	9.84
	F{V}>	4.30
	Y{V}>	3.48
	II>	0.21
	LL>	0.82
	[FY]V>	0.88
	Total	19.06
Non-functional motifs	WW>	0.00
	W{V}>	0.41
	{FY}W>	0.00
	Total	0.41
Candidate amino acids in non-functional position	V{V}>	3.69
	{FY}F>	3.28
	{FY}Y>	1.84
	[IL]{VIL}>	8.29
	{FYIL}[IL]>	11.07

*Standard IUPC one-letter codes for amino acids are used.
†Percentage of a total of 488 human type I membrane proteins from SwissProt and trEMBL databases; non-redundant.

ERGIC-53-related proteins: VIP36 and ERGL

One of the reasons why only a subset of glycoproteins requires ERGIC-53 for efficient ER export may be that other mannose lectins exist that facilitate the transport of other glycoproteins. Moreover, some patients suffering from a combined deficiency of coagulation factors V and VIII were found to have normal expression levels of ERGIC-53, which suggests that there may be redundant cargo receptor systems. The ERGIC-53 homologue VIP36 is such a candidate receptor since, like ERGIC-53, it binds mannose-rich oligosaccharides and cycles in the early secretory pathway [7,34,35]. No specific cargo has been identified for VIP36 *in vivo*, although *in vitro* binding experiments suggest that VIP36 may interact with numerous glycoproteins [34]. There are some notable differences between VIP36 and ERGIC-53: VIP36 lacks the stalk domain of ERGIC-53 as well as the ER dilysine targeting signal (Figure 3). The absence of the ER targeting signal and the longer transmembrane domain may explain why VIP36 is found to localize more to the Golgi apparatus. Unlike ERGIC-53, its main recycling pathway may therefore involve the Golgi apparatus [7,35]. We investigated if VIP36 expression is up-regulated in immortalized lymphocytes from patients suffering from a combined deficiency of factors V and VIII. No such up-regulation was found, either at the mRNA or at the protein level (F. Schnüriger and H.P. Hauri, unpublished work); how-

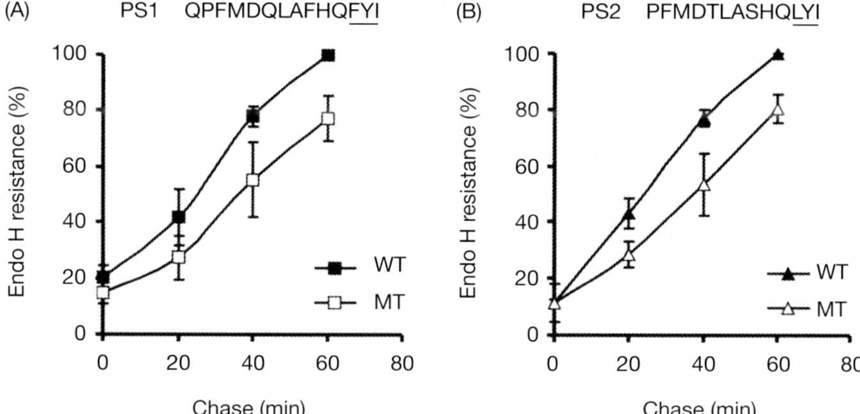

Figure 2 Role of cytoplasmic C-terminal amino acids of presenilins in ER-to-Golgi transport. Amino acid sequences of cytoplasmic tails of (**A**) presenilin 1 (PS1) and (**B**) presenilin 2 (PS2) fused to a CD4 reporter [30] are shown in single letter codes. Underlined residues were substituted with alanine residues in mutant tails. The acquisition of endo H resistance of constructs in pulse/chase experiments is shown. The points represent the means of at least three independent experiments (error bars represent ±S.E.M.). Values obtained after 60 min chase were set to 100%. WT, wild-type tails; MT, mutant tails.

ever, more studies are needed to establish whether or not ERGIC-53 and VIP36 are functionally redundant, complementary or entirely unrelated.

A second ERGIC-53-like gene, ERGL, has been discovered [11]; its mRNA is highly expressed in normal and neoplastic prostate tissue, as well as in the spleen, salivary glands, cardiac atrium and selective cells of the central nervous system. The conceptually translated ERGL shows considerable sequence identity with ERGIC-53 in the N-terminal half and in a more C-terminal segment that includes the transmembrane domain, while the stalk and the cytoplasmic domain are rather different (Figure 3). However, the stalks of both proteins are predicted to adopt an α-helical structure with a coiled-coil domain.

Table 2 Functional cytoplasmic ER-export motifs. Amino acid motifs are shown in standard IUPC one-letter codes. > indicates extreme C-terminal end. TGF-α, transforming growth factor α; MT1-MMP, membrane-type matrix metalloproteinase 1; VSV-G, vesicular stomatitis virus glycoprotein; LAP, lysosomal acid phosphatase. Kir1.1 and 2.1, inward rectifier potassium channel; Sys1p, suppressor of Ypt6 temperature-sensitive mutant.

ER-exit motif	Example(s)
F/YX> and FF	ERGIC-53, VIP36, p24 proteins
FXXXFXXXF	Dopamine D1 receptor
II> and LL>	Emp47p
V>	Pro-TGFα, MT1-MMP, CD8α
DXE	VSV-G, LAP, Kir1.1 and 2.1, Sys1p

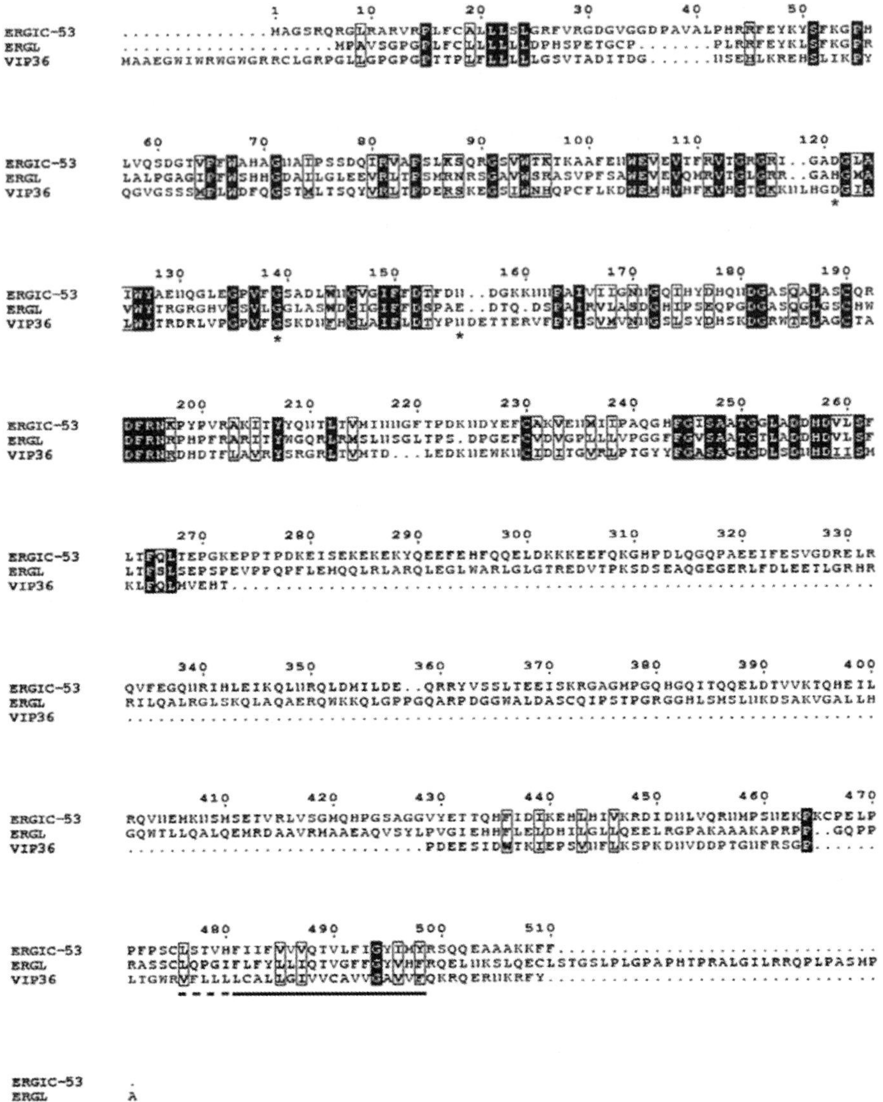

Figure 3 Alignment of ERGIC-53, ERGL and VIP36 amino acid sequences. The sequences of the human proteins were aligned with ClustalW at the EBI server [36]. The amino acid residues are shown in single-letter code and dots indicate gaps. The positions of the amino acids are indicated above the ERGIC-53 sequence. Consensus residues present in all sequences are highlighted in black boxes. Semi-conserved residues are indicated in white boxes. The conserved Asp^{121}, Gly^{139} and Asn^{156} of ERGIC-53 and corresponding residues of the other proteins are marked by asterisks. The transmembrane domain of ERGIC-53 is underlined by a solid line; the elongated membrane domains of ERGL and VIP36 are indicated by a broken line. The accession numbers for GenBank/EMBL (Q) and SwissProt (P) databases are as follows: ERGIC-53, P49257 [37]; ERGL, Q9HAT1 [11]; VIP36, Q12907 (E. Hartmann, B. Reimann, D. Goerlich, T.A. Rapoport and S. Prehn, unpublished work).

Another notable difference is the lack of conservation in ERGL of the residues corresponding to Asp^{121} and Asn^{156} in ERGIC-53. These residues are conserved in plant lectins, VIP36 and ERGIC-53, and are required for mannose binding [8,9]. We have expressed tagged ERGL cDNA in cell culture (L. Liang, I. Pastan and H.-P. Hauri, unpublished work); surprisingly, ERGL was found to be restricted to the ER. It hetero-oligomerizes with ERGIC-53 and blocks its recycling when over-expressed. ERGL may therefore be a regulator of ERGIC-53, but experiments with the untagged protein are required to test this notion.

Conclusion and perspectives

The lectin ERGIC-53 has provided novel information on numerous aspects of the secretory pathway, including the organization of the ER/Golgi system, traffic routes and the mechanism of bidirectional protein trafficking from and to the ER. Despite considerable progress numerous important questions remain unanswered. Do all glycoproteins require assistance by cargo receptor lectins for efficient export from the ER? If they do, additional lectins must exist that are unrelated to ERGIC-53 since the ERGIC-53 lectin family comprises only the three members discussed above. A proteomic approach applied to purified ERGIC membranes may reveal such proteins, since they can be expected to be rather abundant. What is the original function of ERGIC-53 in multicellular organisms? Genetic studies in *C. elegans* can be expected to provide novel insights. What is the precise mechanism by which ERGIC-53 bind and releases its cargo and what determines cargo selectivity? What are the functions of VIP36 and ERGL? Future studies on ERGIC-53 and its homologues may provide further information on the mechanism of selective protein transport in health and in genetic diseases in which exit of key molecules from the early secretory pathway is impaired.

Work described in this chapter was supported by the Swiss National Science Foundation and the University of Basel.

References

1. Hauri, H.P., Appenzeller, C., Kuhn, F. and Nufer, O. (2000) FEBS Lett. **476**, 32–37
2. Ellgaard, L., Molinari, M. and Helenius. A. (1999) Science **296**, 1882–1888
3. Chevet, E., Cameron, P.H., Pelletier, M.F., Thomas, D.Y. and Bergeron, J.J. (2001). Curr. Opin. Struct. Biol. **11**, 120–124
4. Cabral, C.M., Lu, Y. and Sifers, R.N. (2001) Trends Biochem. Sci. **26**, 619–662
5. Michalak, M., Corbett, F.F., Mesaeli, N., Nakamura, K. and Opas, M. (1999) Biochem. J. **344**, 281–292
6. Hauri, H.P., Kappeler, F., Andersson, H. and Appenzeller, C. (2000) J. Cell Sci. **113**, 587–596
7. Dahm, T., White, J., Grill, S., Füllekrug, J. and Stelzer, E.H.K. (2001) Mol. Biol. Cell **12**, 1481–1498
8. Fiedler, K. and Simons, K. (1994) Cell **77**, 625–626
9. Itin, C., Roche, A.C., Monsigny, M. and Hauri, H.P. (1996) Mol. Biol. Cell **7**, 483–493

10. Schröder, S., Schimmoller, F., Singer-Krüger, B. and Riezman, H. (1995) J. Cell Biol. **131**, 895–912
11. Yerusalmi, N., Keppler-Hafkenmeyer, A., Vasmatzis, G., Liu, X.F., Olsson, P., Bera, T.K., Duray, P., Lee, B. and Pastan, I. (2001) Gene **7**, 55–60
12. Schweizer, A., Fransen, J.A., Bächi, T., Ginsel, L. and Hauri, H.P. (1988) J. Cell Biol. **107**, 1643–1653
13. Saraste, J. and Svensson, K. (1991) J. Cell Sci. **100**, 15–30
14. Bannykh, S.I., Rowe, T. and Balch, W.E. (1996) J. Cell Biol. **135**, 19–35
15. Klumperman, J., Schweizer, A., Clausen, H., Tang, B.L., Hong, W., Oorschot, V. and Hauri, H.P. (1998) J. Cell Sci. **111**, 3411–3425
16. Schweizer, A., Fransen, J.A., Matter, K., Kreis, T.E., Ginsel, L. and Hauri, H.P. (1990) Eur. J. Cell Biol. **53**, 85–96
17. Hauri, H.P. and Schweizer, A. (1992) Curr. Opin. Cell Biol. **4**, 600–608
18. Martinez-Menarguez, J.A., Geuze, H.J., Slot, J.W. and Klumperman, J. (1999) Cell **98**, 81–90
19. Stephens, D.J., Lin-Marq, N., Pagano, A., Pepperkok, R. and Paccaud, J.P. (2000) J. Cell Sci. **113**, 2177–2185
20. Hammond, A.T. and Glick, B.S. (2000) Mol. Biol. Cell **11**, 313–330
21. Presley, J.F., Cole, N.B., Schroer, T.A., Hirschberg, K., Zaal, K.J. and Lippincott-Schwartz, J. (1997) Nature (London) **389**, 81–85
22. Scales, S.J., Pepperkok, R. and Kreis, T.E. (1997) Cell **90**, 1137–1148
23. Lippincott-Schwartz, J., Roberts, T.H. and Hirschberg, K. (2000) Annu. Rev. Cell Dev. Biol. **16**, 557–589
24. Vollenweider, F., Kappeler, F., Itin, C. and Hauri, H.P. (1998) J. Cell Biol. **142**, 377–389
25. Appenzeller, C., Andersson, H., Kappeler, F. and Hauri, H.P. (1999) Nat. Cell Biol. **1**, 330–334
26. Moussalli, M., Pipe, S.W., Hauri, H.-P., Nichols, W., Ginsburg, D. and Kaufman, R.J. (1999) J. Biol. Chem. **274**, 32539–32542
27. Nichols, W.C., Seligson, U., Zivelin, A., Terry, V.H., Hertel, C.E., Wheatly, M.A., Moussalli, M.J., Hauri, H.P., Ciavarella, N., Kaufman, R.J. and Ginsburg, D (1998) Cell **93**, 61–70
28. Ouwendijk, J., Moolenaar, C.E., Peters, W.J., Hollenberg, C.P., Ginsel, L.A., Fransen, J.A. and Naim, H.Y. (1996) J. Clin. Invest. **97**, 633–641
29. Kappeler, F., Klopfenstein, D.R., Foguet, M., Paccaud, J.-P. and Hauri, H.P. (1997) J. Biol. Chem. **272**, 31801–31808
30. Nufer, O., Guldbrandsen, S., Degen, M., Kappeler, F., Paccaud, J.P., Tani, K. and Hauri, H.P. (2002) J. Cell Sci. **115**, 619–628
31. Briley, G.P., Hissong, M.A., Chiu, M.L. and Leo, D.C. (1997) Mol. Biol. Cell. **8**, 1619–1631
32. Urena, J.M., Merlos-Suarez, A. Balsenga, J. and Arribas, J. (1999) J. Cell Sci. **112**, 773–784
33. Iodice, L., Sarnataro, S. and Bonatti, S. (2001) J. Biol. Chem. **276**, 28920–28926
34. Hara-Kuge, S., Ohkura, T., Seko, A. and Ymashita, K. (1999) Glycobiology **9**, 833–839
35. Füllekrug, J., Fiedler, K. and Simons, K. (1999) J. Cell Sci. **112**, 2813–2821
36. Thompson, J.D., Higgins, D.G. and Gibson, T.J. (1994) Nucleic Acids Res. **22**, 4673–4680
37. Schindler, R., Itin, C., Zerial, M., Lottspeich, F. and Hauri, H.P. (1993) Eur. J. Cell Biol. **61**, 1–9

Biochem. Soc. Symp. **69**, 83–96
(Printed in Great Britain)

7

New I-type lectins of the CD33-related siglec subgroup identified through genomics

Paul R. Crocker[1] and Jiquan Zhang

Wellcome Trust Biocentre, Division of Cell Biology and Immunology, School of Life Sciences, Dundee University, Dow Street, Dundee DD1 5EH, U.K.

Abstract

Siglecs are sialic-acid-binding proteins of the Ig superfamily that are involved in cell–cell interactions and signalling. In recent years, several novel siglecs that are highly related to CD33/Siglec-2 have been identified through genomics and functional screens. In addition to their distinct sialic-acid-binding properties, most of these novel siglecs bear tyrosine-based signalling motifs that are typically found in inhibitory receptors of the immune system. The restricted expression patterns of CD33-related siglecs in the haemopoietic and immune systems suggests that they are involved in regulating leucocyte activation during inflammatory and immune responses.

Introduction

The siglecs are sialic-acid-binding members of the Ig superfamily [1] that belong to the I-type family of animal lectins [2]. The original members of the siglec family were sialoadhesin (siglec-1), a macrophage adhesion molecule [3], CD22 (siglec-2), a B-cell inhibitory receptor [4], CD33 (siglec-3), a marker of myeloid cells [5] and myelin-associated glycoprotein (MAG; siglec-4), which is expressed on glial cells in the nervous system [6]. These proteins share approx. 25–30% sequence similarity within their extracellular regions. In recent years several novel human, ape and mouse siglec family members have been identified through genomic analyses. All novel siglecs are highly related to CD33 and to each other, with an approx. 50–80% sequence similarity (Figures 1 and 2). They are therefore described collectively as 'CD33-related siglecs' and clearly form a separate subgroup from sialoadhesin, CD22 and MAG, both from functional and evolutionary perspectives. Genes encoding eight genuine human

[1]To whom correspondence should be addressed (e-mail p.r.crocker@dundee.ac.uk).

CD33-related siglecs have been identified and characterized, whereas only five have been identified in the mouse genome [7]. Two of the five murine siglecs still remain to be characterized at the protein level. One siglec-like gene and up to 16 CD33-related siglec pseudogenes have also been identified in humans and two pseudogenes are present in mice [7].

All siglecs are type 1 membrane proteins that possess a characteristic N-terminal sialic acid binding V-set Ig domain, and between one (CD33) and 16 (sialoadhesin) C2 domains that project the sugar-binding site away from the plasma membrane (Figure 1). The cytosolic tails of siglecs are quite variable in sequence and length, although most CD33-related siglecs share regions of high sequence similarity surrounding two conserved tyrosine motifs that are implicated in signalling functions (Figure 2). The genes that encode CD33-related siglecs are clustered on human chromosome 19q13.3–4 or in a syntenic region on mouse chromosome 7p. They appear to have evolved relatively recently by a process involving extensive gene duplication and exon shuffling; however, the

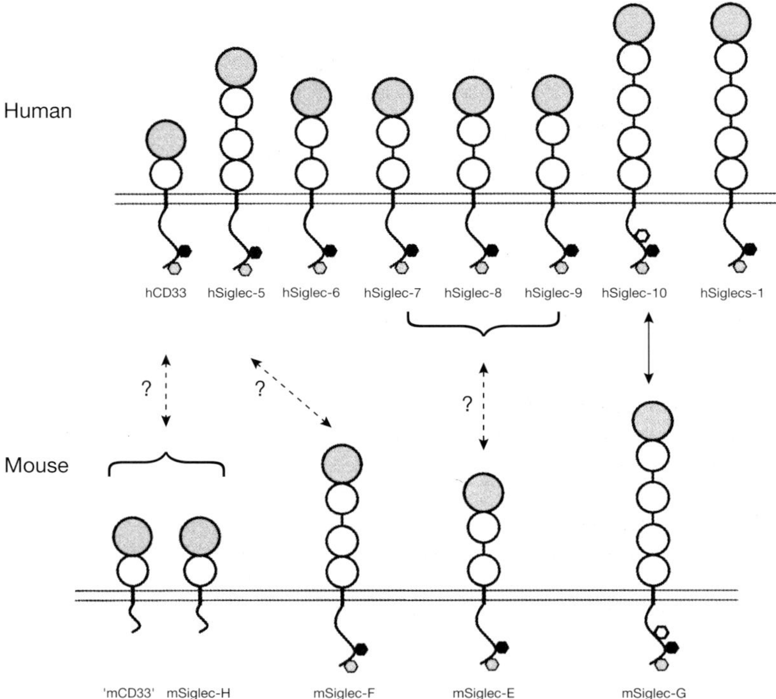

Figure 1 Human and mouse CD33-related siglecs. Full-length forms of all known human and mouse CD33-related siglecs are shown. Individual Ig-like domains are shown as circles. The sialic-acid-binding domain of each protein is shaded. Lines connecting Ig-like domains represent 'linker' regions encoded by separate exons. Hexagons in the cytoplasmic tails represent tyrosine-based motifs. Those shaded in black correspond to immunoreceptor tyrosine-based inhibitory (ITIM) motifs and those in grey correspond to immunoreceptor tyrosine-based switch (ITSM)-like motifs. Potential orthologues are indicated.

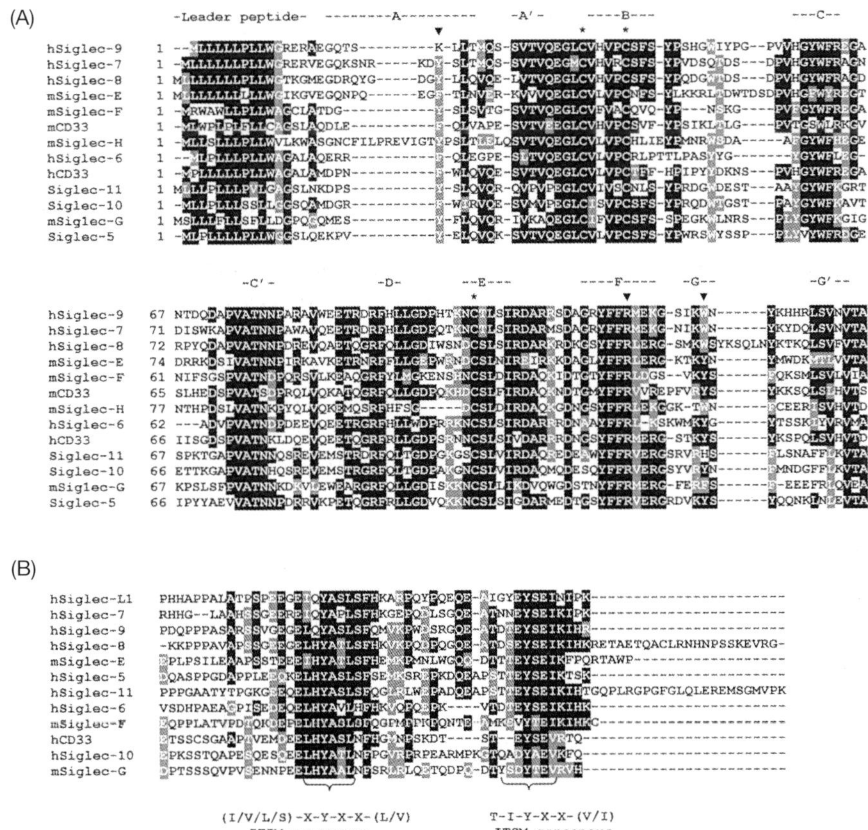

Figure 2 Alignment of sialic-acid-binding V-set domain of all human and mouse CD33-related siglecs (A) and of the cytoplasmic tail C-terminal regions of all human and mouse ITIM-containing CD33-related siglecs (B). (**A**) Residues that are identical in more than 50% of proteins are shaded in black and those that are similar are shaded in grey. The conserved cysteine residues that are typical of siglecs are indicated by asterisks and residues that are important for sialic-acid-dependent binding are indicated by arrowheads. Predicted β-strands (based on the sialoadhesin V-set domain structure) are shown. (**B**) Residues that are identical in more than 50% of proteins are shaded in black and those that are similar are shaded in grey. The positions of the ITIMs and ITSM-like motifs are shown.

repertoire of genes encoding human and mouse CD33-related siglecs appears to be only partially overlapping, which suggests a rapid evolutionary process that differs among mammalian species [7]. This is illustrated by the recent finding that siglec-L1/S2V, a recently-characterized siglec-like protein with two tandem V-set Ig domains, has undergone an inactivating mutation in humans, yet it is still functional in chimpanzees [8]. This chapter focuses on the recently characterized CD33-related siglecs, most of which were identified through genomic studies.

Evolutionary success of the Ig domain and its adaptation to sialic acid recognition

After the initial sequencing of the human genome, approx. 40% of the predicted proteins could be assigned to one of 12 broad families of 'InterPro' entries based on the presence of certain structural domains or motifs. The most populous InterPro entry was found to be the Ig domain, with 765 predicted proteins carrying at least one Ig domain [9]. This was in striking contrast with *Drosophila melanogaster* and *Caenorhabditis elegans* whose proteomes contain far fewer predicted proteins with Ig domains. The dramatic increase in Ig domain-containing proteins during animal evolution correlates with a parallel increase in the number of proteins associated with extracellular recognition events, especially those involved in host immunity [9].

The Ig domain contains two anti-parallel β-sheets that are usually linked by a characteristic disulphide bond and stabilized by a hydrophobic interior. Its evolutionary success in extracellular recognition events is likely to be due to several factors. It has been suggested that Ig domains can display a diversity of sequences that is greater than can be achieved with other protein superfamilies [10]. Sequence diversity can be displayed both on the external faces of the β-sheets as well as on the loops that connect the β-strands. In addition, Ig domains are often encoded by single exons with phase-I splice junctions that allow exon duplication and domain shuffling, leading to gene expansion and additional diversity. The Ig domain is also inherently resistant to proteolytic attack, and is therefore able to withstand the potentially harsh conditions in the extracellular milieu.

It is well known that V-set domains of many Ig molecules can bind to diverse carbohydrate structures including sialic acid; however, it is very unlikely that the ancestral V-set domain of siglecs evolved from an Ig gene. While Igs bind carbohydrate antigens via residues that are located on the hypervariable loop regions, siglecs bind sialylated glycans using residues located on the A, F and G β-strands in a manner that is more similar to that seen for other members of the Ig superfamily involved in protein–protein interactions [11–13]. Interestingly, one of the closest structural matches to the Sn V-set domain is the P0 adhesion molecule that is found in myelin. P0 is a single V-set-containing transmembrane protein that has been proposed to represent a typical primordial Ig-like domain involved in homophilic protein–protein interactions [10,14]. It is therefore likely that the siglec family evolved from an ancestral 'P0-like' gene and then, through mutation and selection, acquired the ability to interact specifically with sialic acids. A similar process could have occurred independently for the P0-related protein CD83 expressed by dendritic cells [15]. This receptor has a single V-set domain that can bind cells in a sialic-acid-dependent manner [15] yet has none of the typical features of siglec V-set domains.

If the above scenario is correct, siglecs would have appeared subsequently to the evolution of sialic acid biosynthesis pathways. Sialic acids are present at high levels in deuterostome-lineage animals such as starfish, but are absent or expressed at low levels in the protostome-lineage species *C. elegans* and *D. melanogaster* [16]. Genome-wide searches of *C. elegans* and *D. melanogaster* failed to uncover obvious siglec homologues or sequences encoding known enzymes of the sialic acid biosynthesis pathway [16]; it is not known at present if siglecs are present in species such as starfish. MAG (siglec-4) has been identified immunochemically in a wide variety of higher non-mammalian species including fish, frogs, birds, snakes and lizards [17,18] and 'expressed sequence tag' (EST) database searches indicate the presence of transcribed siglec-like sequences in the African clawed frog *Xenopus laevis* and the zebrafish *Danio rerio* [7] (P.R. Crocker, unpublished work). The future availability of complete genome projects for both vertebrate and invertebrate species will be of great interest in understanding the evolutionary pathways of siglecs.

The V-set domains of siglecs are particularly well adapted for sialic acid recognition. Long before they were known to be carbohydrate-binding proteins, Williams et al. [10] pointed out that MAG and CD33 (the first members of the siglec family to be cloned and sequenced) had an unusual arrangement of three cysteine residues in the N-terminal V-set Ig domain. These are completely conserved in all siglecs (Figure 2A) and were predicted [14] to give rise to a disulphide bond within the ABED β-sheet rather than between the ABED and GFCC′ β-sheets. The crystal structure of the V-set domain of sialoadhesin complexed with 3′ sialyllactose confirmed this arrangment [13]. Moreover, it showed how the intrasheet disulphide bond resulted in an increased separation between the β-sheets and exposure of two normally buried tryptophan residues on β-strands A and G [13]. These residues make important hydrophobic contacts with the N-acetyl and glycerol moieties of the sialic acid molecule. In addition, a highly conserved arginine residue on the F β-strand (Figure 2A) forms a critical salt bridge with the carboxylate group of sialic acid [13]. This template for sialic acid recognition is shared by all known siglecs and provides a platform onto which additional specificity for sialic acid types and linkages to adjacent sugars can be built. The variable C–C′ loop (Figure 2A), for example, may be important for determining fine specificity, as demonstrated recently for the human siglecs hSiglec-7 and hSiglec-9 [19].

Naturally occurring mutations in the critical arginine residue have been found in a human siglec-like gene that is designated siglec-L1/S2V [8,20]. This unusual molecule contains two tandem V-set domains with Arg→Cys and Arg→Gln mutations. Remarkably, approx. 50% of the human and mouse siglec pseudogenes have mutations in the codon that encodes the critical arginine residue [7]. Phylogenetic analyses suggest that these are likely to have been independent events that could have functionally inactivated the proteins [7]. A lack of positive selection pressure could then lead to additional mutations and eventual gene inactivation. If correct, this interpretation would support the idea that sialic acid recognition is essential in mediating the biological functions of CD33-related siglecs and maintaining them as functional genes within populations.

Identification of novel siglecs through genomics

The creation of extensive databases containing random sequences from cDNA and genomic clones, together with powerful search engines, has enabled facile detection of new gene family members. hSiglec-5, -7, -8 and -10 were identified as CD33-related sequences in the Human Genome Science's EST database, which contains more than 1 million ESTs derived from approx. 700 cDNA libraries [21–25]. hSiglec-10 was also obtained from random sequencing of a dendritic cell cDNA library [26], from public genome databases [27] and from searches of Incyte's EST database which is derived from 995 libraries with more than 4 million clones [28]. Genomic and EST databases were used to identify hSiglec-6, -7, -9 and -11 [16,29–33], as well as the murine siglecs, mSiglec-E, -F, -G and -H [7,34] (P.R. Crocker and J. Zhang, unpublished work). In all cases, examination of the predicted open reading frames revealed the presence of the conserved amino acid residues that are important for sialic acid recognition and, where tested, sialic acid binding has also been demonstrated experimentally. Several CD33-related sequences were identified in functional screens. Siglec-6 cDNA was isolated as a low-affinity leptin-binding protein in an expression cloning strategy and was shown subsequently to bind sialic acids [35]. Siglec-7 was the target in a screen for antigens on human natural killer (NK) cells that modulate killing activity and designated p75/AIRM1 [36]. mSiglec-E was isolated in a yeast two-hybrid screen using the protein tyrosine phosphatase Src homology 2-domain-containing protein tyrosine phosphatase-1 (SHP-1) as bait [37]. Finally, hSiglec-8, -9, -10, -L1 and mSiglec-E, -F and -H were identified by the combined approaches of PCR and low-stringency cDNA/genomic library screening (J. Zhang and P.R. Crocker, unpublished work).

The human siglec-like sequence hSiglec-L1, which is discussed above, was identified in public genomic and EST databases [8,38]. The same gene was isolated in a yeast two-hybrid screen using SHP-1 as a bait and was, surprisingly, later shown to bind to red blood cells in a sialic-acid-dependent manner when expressed in COS cells [20]. However, this result is controversial, especially in light of the findings that the chimpanzee orthologue of hSiglec-L1 has retained the critical arginine in the N-terminal V-set domain and mediates robust sialic-acid-dependent binding, in stark contrast with hSiglec-L1 [8]. Further genomic analyses in other species are required to shed light on the evolutionary pathways taken by CD33-related siglecs.

Comparisons of human and mouse CD33-related siglecs

Recent access to the almost complete human and murine (Celera Genomics) genomic databases has allowed the precise cataloguing and mapping of all human and mouse CD33-related genes [7,39]. While the human genome contains eight CD33-related siglecs, one siglec-like gene and approx. 16 pseudogenes, the mouse genome contains only five CD33-related genes and two pseudogenes [7,33]. All eight of the human siglecs have two conserved immunoreceptor tyrosine-based inhibitory motifs (ITIMs) in their cytoplasmic tails, which are implicated in inhibitory signalling functions (see below). In contrast, only three of the five murine CD33-related genes encode ITIM-containing siglecs and those that lack ITIMs do not have obvious orthologues in humans (Figure 1 and Table 1).

Consistent with the earlier *in situ* hybridization work carried out on hSiglec-5, -6, -7, -8 and -10, these genes are clustered within a 506-kb region on human chromosome 19q13.3–13.4. hSiglec-11 is located approx. 1 Mb upstream from the cluster, but still within the cytological band 19q13.3–13.4 [33]. Apart from mSiglec-H, all murine CD33-related siglecs lie in a syntenic region on chromosome 7p. Chromosome 19 in humans is the most gene-rich chromosome; it contains large families of Ig-related receptors, including the carcinoembryonic antigen family, killer inhibitory receptors and Ig-like transcripts, in addition to the CD33-related siglecs (reviewed in [40]). The dramatic expansion of Ig-like families on chromosome 19q may be due, at least in part, to the presence of chromosome 19-specific minisatellites that facilitated gene duplication by acting as nucleation sites for unequal cross-overs during meiotic recombination [16,41]. Such exchanges can also introduce nucleotide changes in the non-coding regulatory regions, which, in the case of siglecs, could be important for their acquiring cell-type-specific expression patterns in the immune system.

The sequence similarity shared between all of the CD33-related siglecs in humans and mice (approx. 50–80%) falls generally within the range for orthologous pairs of immune system receptors. This makes assignments of orthologues potentially problematic. Indeed, in sequence alignments and phylogenetic analysis of all known human and mouse siglecs, only siglec-1, -2, -4 and -10 can be seen to have clear orthologues in both species. Where studied, the corresponding proteins in humans and mice exhibit similar sialic-acid-binding properties and cellular expression patterns [3,42–44] (P.R. Crocker and J. Zhang, unpublished work). Based on gene structure, sequence alignments, relative chromosomal localization and, where available, functional properties, it has been suggested that murine CD33, mSiglec-E and mSiglec-F are orthologues of human CD33, hSiglec-9 and hSiglec-5 respectively [7]. However, murine CD33 has a completely different cytoplasmic tail from human CD33 and it lacks ITIM-like motifs [5,45]. Unlike mSiglec-G and hSiglec-10, which are approx. 60% identical, mSiglec-E and -F are only approx. 50% identical with hSiglec-9 and -5 respectively (Table 1) and show significant differences in sialic-acid-binding specificity and cellular expression patterns [7] (P.R. Crocker and J. Zhang, unpublished work). Therefore, based on currently available data, it is possible

Table 1 Percentage sequence identities derived from alignments of the indicated human and mouse siglecs. Alignments were made using the full-length protein sequences, except for those made using human CD33, for which only the Ig domains were used.

Human	Mouse	Ig domains	Identity
MAG	MAG	5	94%
Sialoadhesin	Sialoadhesin	17	73%
CD22	CD22	7	59%
CD33	'CD33'	2	61%
CD33	Siglec-H	2	61%
Siglec-5	Siglec-F	4	49%
Siglec-7	Siglec-E	3	53%
Siglec-8	Siglec-E	3	54%
Siglec-9	Siglec-E	3	53%
Siglec-10	Siglec-G	5	60%

that siglec-1, -2, -4 and -10 were the principal siglecs present in the common ancestor of humans and mice. The remaining CD33-related siglecs could have evolved independently in different species, involving a process of exon deletion and gene duplication, starting from a siglec-10-like precursor. This would have been particularly extensive in the primate lineages, leading to the generation of eight functional CD33-related siglecs and one siglec-like molecule.

Sialic-acid-binding specificity

Sialic acid is a generic term for a large family of 9-carbon sugars that are all derivatives of neuraminic acid or oxo-deoxynonulosonic acid. They are typically found at the exposed, non-reducing ends of oligosaccharide chains attached to a wide variety of proteins and lipids. At least 18 sialyltransferases exist in mammals and these catalyse the transfer of different types of sialic acid to various acceptors, to generate α2,3-, α2,6- or α2,8-glycosidic linkages. Despite the high degree of sequence similarity, the CD33-related siglecs exhibit clear differences in their sialic-acid-binding preferences, both at the level of apparent affinity and preference for glycosidic linkage. hSiglec-7, for example, shows a striking preference for α2,8-linked disialic acids in contrast with the highly related hSiglec-9 which strongly prefers α2,3-linkages [19]. The biological significance of these sugar-binding preferences is unclear at present and it is likely to require an understanding of the precise functions of each molecule, about which very little is known at present.

With the exception of hSiglec-6, which shows restricted specificity for the sialyl Tn antigen [35], all of the known siglecs recognize forms and linkages of sialic acid that are commonly found at cell surfaces and in the extracellular environment [7,16,21–23,25,30,31,33,43]. One consequence is that the siglec-binding sites can be masked by *cis* interactions with sialic acids on the same cell, thereby preventing them from mediating cell–cell interactions. Since all CD33-related siglecs on circulating leucocytes appear to be masked [46], one possibility is that the binding specificity of each siglec is tailored towards the sialylation pattern of the host cell, thereby facilitating *cis* interactions with relevant ligands that play a role in modulation of cellular functions. The importance of such sialic-acid-dependent *cis* interactions has been demonstrated recently for CD22/siglec-2, which associates with the B-cell receptor and inhibits B-cell activation [47]. CD22 specifically binds α2,6-linked sialic acids on N-glycans that are generated by the ST6GalI sialyltransferase, an enzyme that is expressed abundantly by β1,4-galactoside α2,6 sialytransferase [48]. Addition of a novel CD22-specific sugar inhibitor to B-cells was found to enhance B-cell activation, thereby showing that sialic-acid-dependent *cis* interactions are important for the signalling functions of CD22 [49].

Expression pattern of CD33-related siglecs

CD33 was originally identified as a marker of myeloid progenitor cells that is absent from multipotential stem cells in the bone marrow [50,51]. This restricted specificity has been extremely useful in providing a means of distinguishing subpopulations of haemopoietic progenitors in samples of bone marrow and umbilical cord blood. Furthermore, its expression on the great majority of acute myeloid leukaemic blasts has been exploited in the development of a humanized antibody conjugated to the toxin calicheamicin (Mylotarg®), to eliminate residual leukaemia cells after relapse (reviewed in [52]). Recent studies have shown that, similar to CD33 and other siglecs, the novel CD33-related siglecs are expressed in a very cell-type-specific manner, largely within the haemopoietic system. While much remains to be learned, some intriguing observations have emerged from studies of normal human circulating blood leucocytes and selected tissues using panels of monoclonal antibodies. Some siglecs are expressed quite broadly, e.g. hSiglec-9 is found on neutrophils, monocytes and on a substantial fraction of NK cells and B-cells [30]. Others are much more restricted in their expression, notably hSiglec-8, which is found only on circulating eosinophils and at very low levels on basophils [24,53]. On the other hand, several siglecs can be present on the same cell type; for example, monocytes express hCD33, hSiglec-5, -7, -9 and –10, and circulating B-cells express hSiglec-5, -6, -9 and -10, suggesting some degree of functional redundancy at the cellular level [21,22,25,30,35,54]. Siglec-11 is the only CD33-related siglec characterized thus far that is undetectable on blood leucocytes [33]. Interestingly, siglec-11 is expressed by certain tissue macrophages, including brain microglia and liver Kupffer cells and certain other leucocytes, especially in inflammation [33]. There is only limited information available at present on the expression of the other CD33-related siglecs on cells outside of the haemopoietic system, although hSiglec-6 is expressed at high levels by cytotrophoblasts and syncytiotrophoblasts of the placenta, in addition to B-cells [35]. Interestingly, using a chicken antibody, siglec-L1 was observed at high levels on the apical edge of epithelial cells in various tissues [8]. However, other studies have shown a pattern of siglec-L1 mRNA expression that is more consistent with haemopoietic expression, especially on macrophages [20,38] (J. Zhang and P.R.Crocker, unpublished work).

CD33-related siglecs as inhibitory receptors

It is now well established that balancing positive and negative signals is crucial for setting activation thresholds for cells of the haemopoietic and immune systems. The paradigm that has emerged from studies of classical inhibitory receptors such as the FcγRIIb and killer inhibitory receptors is that when cellular activation is triggered by receptors with immunoreceptor tyrosine-based activation motifs (ITAMs), counteracting inhibitory signals are delivered through receptors bearing ITIMs (reviewed in [55]). Tyrosine-phosphorylated ITIMs recruit and activate cytosolic phosphatases, either the protein tyrosine phosphatases Src homology 2-domain-containing protein tyrosine phosphatase (SHP)-1 or SHP-2, or the inositol polyphosphate 5′ phosphatase SHIP (Src homology 2-containing inositol-polyphosphate 5′ phosphatase). These phosphatases inhibit signalling pathways by distinct mechanisms, which results in raised activation thresholds (reviewed in [55]).

The presence of two conserved ITIM-like motifs in the cytoplasmic regions of all hCD33-related siglecs (Figure 2B) and the differential expression of these proteins on different types of leucocyte, suggest a generic role in regulating cellular activation. hSiglec-8, which was originally thought to lack ITIM-like motifs, has now been found to exist in alternatively spliced forms containing either a short cytoplasmic tail without ITIMs or a long tail with ITIMs [23–25,56,57]. While the membrane proximal motif of CD33-related siglecs fits the consensus ITIM sequence (Ile/Val/Leu/Ser)-Xaa-Tyr-Xaa-Xaa-(Leu/Val), the membrane distal motif does not (Figure 2B). Detailed studies of hCD33, hSiglec-10, hSiglec-L1 and mouse inhibitory siglec (MIS)/mSiglec-E have shown that both SHP-1 and SHP-2 can be recruited following pervanadate treatment of cells to inhibit tyrosine phosphatases. Mutagenesis studies have shown that for all of the above siglecs the distal ITIM-like motif binds more weakly to SHP-1 and SHP-2, and is dispensable for phosphatase binding, presumably reflecting its departure from the consensus ITIM sequence. However, the distal motif is highly conserved (Figure 2B), which suggests that it is important for interacting with other regulatory molecules. Interestingly, this motif is similar to a tyrosine-based motif [Thr-Xaa-Tyr-Xaa-Xaa-(Val/Ile)] found in SLAM (signalling lymphocyte activation molecule) and SLAM-related proteins that binds SAP (SLAM-associated protein), a single-SH2-domain protein that inhibits SHP-2 recruitment and activates additional signalling pathways (reviewed in [58]). This motif has recently been designated 'immunoreceptor tyrosine-based switch motif' (ITSM), which reflects its apparent role in switching between different signalling pathways [59].

Functional evidence indicating that siglecs can mediate inhibitory signals has been obtained using antibodies to co-cross-link hCD33 or MIS/mSiglec-E with an activating human receptor, FcγR1 [34,60,61]. This resulted in reduced Ca^{2+} influx compared with cross-linking FcγRI alone. Likewise, siglec-7 was identified as an inhibitory NK cell receptor in a redirected killing assay in which anti-siglec-7 antibodies were used to cluster siglec-7 at the NK cell–target-cell interface [54]. In other functional studies, the addition of intact anti-hCD33 or anti-hSiglec-7 monoclonal antibodies to haemopoietic cell cultures resulted in reduced cell growth and prevention of dendritic cell development [36,62]. The physiological relevance of these interesting findings remains unclear, since antibodies rather than natural ligands were used to cluster the siglecs. Clearly, additional experimental approaches are needed to understand the functions of CD33-related siglecs and the importance of sialic acid recognition in mediating these functions.

Research in the authors' laboratory is supported by the Wellcome Trust and Biotechnology and Biological Sciences Research Council. We are grateful to Helen Attrill for advice on sequence alignments.

References

1. Crocker, P.R., Clark, E.A., Filbin, M., Gordon, S., Jones, Y., Kehrl, J.H., Kelm, S., Le Douarin, N., Powell, L., Roder, J. et al. (1998) Glycobiology **8**, v
2. Powell, L.D. and Varki, A. (1995) J. Biol. Chem. **270**, 14243–14246

3. Crocker, P.R., Mucklow, S., Bouckson, V., McWilliam, A., Willis, A.C., Gordon, S., Milon, G., Kelm, S. and Bradfield, P. (1994) EMBO J. **13**, 4490–4503
4. Stamenkovic, I. and Seed, B. (1990) Nature (London) **345**, 74–77
5. Simmons, D.L. and Seed, B. (1988) J. Immunol. **141**, 2797–2800
6. Arquint, M., Roder, J., Chia, L.S., Down, J., Wilkinson, D., Bayley, H., Braun, P. and Dunn, R. (1987) Proc. Natl. Acad. Sci. U.S.A. **84**, 600–604
7. Angata, T., Hingorani, R., Varki, N.M. and Varki, A. (2001) J. Biol. Chem. **276**, 45128–45136
8. Angata, T., Varki, N.M. and Varki, A. (2001) J. Biol. Chem. **276**, 40282–40287
9. Lander, E.S., Linton, L.M., Birren, B., Nusbaum, C., Zody, M.C., Baldwin, J., Devon, K., Dewar, K., Doyle, M., FitzHugh, W. et al. (2001) Nature (London) **409**, 860–921
10. Williams, A.F., Davis, S.J., He, Q. and Barclay, A.N. (1989) Cold Spring Harb. Symp. Quant. Biol. **54**, 637–647
11. Vinson, M., van der Merwe, P.A., Kelm, S., May, A., Jones, E.Y. and Crocker, P.R. (1996) J. Biol. Chem. **271**, 9267–9272
12. van der Merwe, P.A., Crocker, P.R., Vinson, M., Barclay, A.N., Schauer, R. and Kelm, S. (1996) J. Biol. Chem. **271**, 9273–9280
13. May, A.P., Robinson, R.C., Vinson, M., Crocker, P.R. and Jones, E.Y. (1998) Mol. Cell **1**, 719–728
14. Williams, A.F. and Barclay, A.N. (1988) Annu. Rev. Immunol. **6**, 381–405
15. Scholler, N., Hayden-Ledbetter, M., Hellstrom, K.E., Hellstrom, I. and Ledbetter, J.A. (2001) J. Immunol. **166**, 3865–3872
16. Angata, T. and Varki, A. (2000) J. Biol. Chem. **275**, 22127–22135
17. O'Shannessy, D.J., Willison, H.J., Inuzuka, T., Dobersen, M.J. and Quarles, R.H. (1985) J. Neuroimmunol. **9**, 255–268
18. Tropak, M.B., Jansz, G.F., Abramow-Newerly, W. and Roder, J.C. (1995) Comp. Biochem. Physiol. B. Biochem. Mol. Biol. **112**, 345–354
19. Yamaji, T., Teranishi, T., Alphey, M.S., Crocker, P.R. and Hashimoto, Y. (2002) J. Biol. Chem. **277**, 6324–6332
20. Yu, Z., Lai, C.M., Maoui, M., Banville, D. and Shen, S.H. (2001) J. Biol. Chem. **276**, 23816–23824
21. Cornish, A.L., Freeman, S., Forbes, G., Ni, J., Zhang, M., Cepeda, M., Gentz, R., Augustus, M., Carter, K.C. and Crocker, P.R. (1998) Blood **92**, 2123–2132
22. Nicoll, G., Ni, J., Liu, D., Klenerman, P., Munday, J., Dubock, S., Mattei, M.G. and Crocker, P.R. (1999) J. Biol. Chem. **274**, 34089–34095
23. Floyd, H., Ni, J., Cornish, A.L., Zeng, Z., Liu, D., Carter, K.C., Steel, J. and Crocker, P.R. (2000) J. Biol. Chem. **275**, 861–866
24. Kikly, K.K., Bochner, B.S., Freeman, S.D., Tan, K.B., Gallagher, K.T., D'Alessio, K.J., Holmes, S.D., Abrahamson, J.A., Erickson-Miller, C.L., Murdock, P.R. et al. (2000) J. Allergy Clin. Immunol. **105**, 1093–1100
25. Munday, J., Kerr, S., Ni, J., Cornish, A.L., Zhang, J.Q., Nicoll, G., Floyd, H., Mattei, M.G., Moore, P., Liu, D. and Crocker, P.R. (2001) Biochem. J. **355**, 489–497
26. Li, N., Zhang, W., Wan, T., Zhang, J., Chen, T., Yu, Y., Wang, J. and Cao, X. (2001) J. Biol. Chem. **276**, 28106–28112
27. Yousef, G.M., Ordon, M.H., Foussias, G. and Diamandis, E.P. (2001) Biochem. Biophys. Res. Commun. **284**, 900–910
28. Whitney, G., Wang, S., Chang, H., Cheng, K.Y., Lu, P., Zhou, X.D., Yang, W.P., McKinnon, M. and Longphre, M. (2001) Eur. J. Biochem. **268**, 6083–6096
29. Takei, Y., Sasaki, S., Fujiwara, T., Takahashi, E., Muto, T. and Nakamura, Y. (1997). Cytogenet. Cell. Genet. **78**, 295–300
30. Zhang, J.Q., Nicoll, G., Jones, C. and Crocker, P.R. (2000) J. Biol. Chem. **275**, 22121–22126
31. Angata, T. and Varki, A. (2000) Glycobiology **10**, 431–438

32. Foussias, G., Yousef, G.M. and Diamandis, E.P. (2000) Genomics **67**, 171–178
33. Angata, T., Kerr, S.C., Greaves, D.R., Varki, N.M., Crocker, P.R. and Varki, A. (2002) J. Biol. Chem., **277**, 24466–24474
34. Ulyanova, T., Shah, D.D. and Thomas, M.L. (2001) J. Biol. Chem. **276**, 14451–14458
35. Patel, N., Linden, E.C., Altmann, S.W., Gish, K., Balasubramanian, S., Timans, J.C., Peterson, D., Bell, M.P., Bazan, J.F., Varki, A. and Kastelein, R.A. (1999) J. Biol. Chem. **274**, 22729–22738
36. Vitale, C., Romagnani, C., Falco, M., Ponte, M., Vitale, M., Moretta, A., Bacigalupo, A., Moretta, L. and Mingari, M.C. (1999) Proc. Natl. Acad. Sci. U.S.A. **96**, 15091–15096
37. Yu, Z., Maoui, M., Wu, L., Banville, D. and Shen, S.-H. (2001) Biochem. J. **353**, 483–492
38. Foussias, G., Taylor, S.M., Yousef, G.M., Tropak, M.B., Ordon, M.H. and Diamandis, E.P. (2001) Biochem. Biophys. Res. Commun. **284**, 887–899
39. Yousef, G.M., Ordon, M.H., Foussias, G. and Diamandis, E.P. (2002) Gene **286**, 259–270
40. Trowsdale, J., Barten, R., Haude, A., Stewart, C.A., Beck, S. and Wilson, M.J. (2001) Immunol. Rev. **181**, 20–38
41. Fitch, D.H., Bailey, W.J., Tagle, D.A., Goodman, M., Sieu, L. and Slightom, J.L. (1991) Proc. Natl Acad. Sci. U.S.A. **88**, 7396–7400
42. Powell, L.D., Sgroi, D., Sjoberg, E.R., Stamenkovic, I. and Varki, A. (1993) J. Biol. Chem. **268**, 7019–7027
43. Kelm, S., Pelz, A., Schauer, R., Filbin, M.T., Tang, S., de Bellard, M.E., Schnaar, R.L., Mahoney, J.A., Hartnell, A., Bradfield, P. and Crocker, P.R. (1994) Curr. Biol. **4**, 965–972
44. Hartnell, A., Steel, J., Turley, H., Jones, M., Jackson, D.G. and Crocker, P.R. (2001) Blood **97**, 288–296
45. Tchilian, E.Z., Beverley, P.C., Young, B.D. and Watt, S.M. (1994) Blood **83**, 3188–3198
46. Razi, N. and Varki, A. (1999) Glycobiology **9**, 1225–1234
47. Cyster, J.G. and Goodnow, C.C. (1997) Immunity **6**, 509–517
48. Hennet, T., Chui, D., Paulson, J.C. and Marth, J.D. (1998) Proc. Natl. Acad. Sci. U.S.A. **95**, 4504–4509
49. Kelm, S., Brossmer, R., Gerlach, J., Danzer, C.-P. and Nitschke, L. (2002) J. Exp. Med. **195**, 1207–1213
50. Griffin, J.D., Linch, D., Sabbath, K., Larcom, P. and Schlossman, S.F. (1984) Leuk. Res. **8**, 521–534
51. Andrews, R.G., Torok-Storb, B. and Bernstein, I.D. (1983) Blood **62**, 124–132
52. Bernstein, I.D. (2002) Clin. Lymphoma **2** (Suppl. 1), S9–S11
53. Floyd, H., Nitschke, L. and P.R.Crocker. (2000) Immunology **101**, 342–347
54. Falco, M., Biassoni, R., Bottino, C., Vitale, M., Sivori, S., Augugliaro, R., Moretta, L. and Moretta, A. (1999) J. Exp. Med. **190**, 793–802
55. Ravetch, J.V. and Lanier, L.L. (2000) Science **290**, 84–89
56. Foussias, G., Yousef, G.M. and Diamandis, E.P. (2000) Biochem. Biophys. Res. Commun. **278**, 775–781
57. Aizawa, H., Plitt, J. and Bochner, B.S. (2002) J. Allergy Clin. Immunol. **109**, 176
58. Veillette, A. (2002) Sci. Signal Transduction Knowledge Environ. **2002**, E8
59. Shlapatska, L.M., Mikhalap, S.V., Berdova, A.G., Zelensky, O.M., Yun, T.J., Nichols, K.E., Clark, E.A. and Sidorenko, S.P. (2001) J. Immunol. **166**, 5480–5487
60. Ulyanova, T., Blasioli, J., Woodford-Thomas, T.A. and Thomas, M.L. (1999) Eur. J. Immunol. **29**, 3440–3449
61. Paul, S.P., Taylor, L.S., Stansbury, E.K. and McVicar, D.W. (2000) Blood **96**, 483–490
62. Ferlazzo, G., Spaggiari, G.M., Semino, C., Melioli, G. and Moretta, L. (2000) Eur. J. Immunol. **30**, 827–833

Biochem. Soc. Symp. **69**, 95–103
(Printed in Great Britain)

8

Roles of galectins *in vivo*

Françoise Poirier[1]

Institut Jacques Monod, 2 Place Jussieu, 75251 Paris cedex 05, France

Abstract

The mutational analysis of the galectin family is shedding a different light of this class of molecules. On the one hand, it appears that galectin 1 and galectin 3 are not required for the survival of mice in normal animal house conditions, while on the other hand, there seems to be several subtle, but very complex, consequences of lacking galectins during development.

Introduction

Galectins, or S-type lectins, are a family of carbohydrates-recognition proteins that are defined by their shared structure and their affinity for β-galactoside-containing glycoconjugates [1–3]. Galectins were originally isolated from mammalian tissues, but they have now been found in many other species including birds, fish, worms and sponges [4]. Twelve different mammalian galectins have been identified to date. Despite the absence of signal peptides, galectins can be secreted outside the cells by a novel pathway independent of the Golgi apparatus [5]. Members of the galectin family have been found in all subcellular compartments (extracellular, cytoplasmic and nuclear) *in vivo*, depending on the tissue type and the stage of the cell cycle. The subcellular localization is likely to be a crucial parameter in the understanding of the function fulfilled by a given galectin at a given time.

Most of the work on galectins concerns their potential roles in tumour biology and inflammation [6,7]. In addition, galectins have been studied in many *in vitro* tissue culture systems, but no simple unifying theory is emerging from this wealth of data. Instead, several different functions have been assigned to galectins, ranging from structural roles to signal transmission; they have been described as modulators of cell adhesion [8–11], regulators of cell survival [12,13] and even as splicing factors [14].

With the goal of identifying developmental processes in which galectins participate *in vivo*, our laboratory has initiated a genetic analysis of the galectin gene family in the mouse. Mouse lines carrying deletions of the galectin 1 or

[1]E-mail poirier@ijm.jussieu.fr

galectin 3 genes have been generated. Because both mutations are viable and fertile, these animal models provide good tools to unravel the various roles of this class of molecules *in vivo*.

Different galectins are expressed in different tissues during development

We first examined the distribution of galectin gene transcripts by *in situ* hybridization and, in some cases, by antibody staining on whole mouse embryo sections. This analysis has thus far been carried out for galectins 1, 3, 4/6 and 7 (Figure 1). All these genes are expressed during development, and each of them displays a restricted and unique expression pattern. These characteristics are also shared by galectin 5 even though it has not yet been characterized as extensively. Galectin 1 first appears in the trophectoderm (extra-embryonic) cells of the blastocyst at the time of implantation (E4.5, day 4.5 of embryogenesis), and starting from E9.5, galectin 1 becomes broadly distributed in the embryo proper, although this distribution is restricted mostly to tissues of mesodermic origin [15].

Galectin 3 is also first detected in the trophectoderm cells at E4.5. Within the embryo, its expression is confined almost entirely to developing bone tissues until midgestation. At later stages, galectin 3 transcripts are also found in various epithelial tissues, including kidney, bladder, intestine and skin [16]. In addition, galectin 3, which is also known as Mac2 [17], is expressed in the macrophage lineage.

Galectin 4 and galectin 6 are both found exclusively in the epithelial cells of the developing intestine [18]; however, given the extensive degree of sequence similarity between galectin 4 and galectin 6, the *in situ* results had to be corroborated by RNase protection experiments. Finally, galectin 7 expression is restricted to stratified epithelia, regardless of their state of keratinization or their regional specialization. Therefore, the main galectin 7-expressing tissue is the epidermis, but it is also found in the oesophagus, tongue, lip, cornea and Sertoli cells [19].

Several conclusions can be drawn from these observations: (i) all tested galectins are expressed in the developing embryo; (ii) each appears to be involved in specific developmental processes, e.g. galectin 3 in bone formation, galectins 4 and 6 in intestine morphogenesis, and galectin 7 in epithelial stratification; (iii) apart from galectins 4 and 6, which clearly result from a recent gene duplication [20], so far, no two galectins are expressed in the same way. Although galectin 1 and galectin 3 transcripts, for example, are simultaneously present in the trophectoderm cells of the implanting embryo, the corresponding proteins have strikingly different subcellular localization; galectin 1 is found mainly in the cytoplasm, while galectin 3 is nucleus- and membrane-associated [21]. Similarly, galectins 3 and 7, are co-expressed in the epidermis, but galectin 3 behaves like all other suprabasal differentiation markers, whereas galectin 7 displays a totally novel expression pattern, in the sense that it is barely detectable when there is a single layer of suprabasal cells and is very strongly expressed when there are several layers of suprabasal cells.

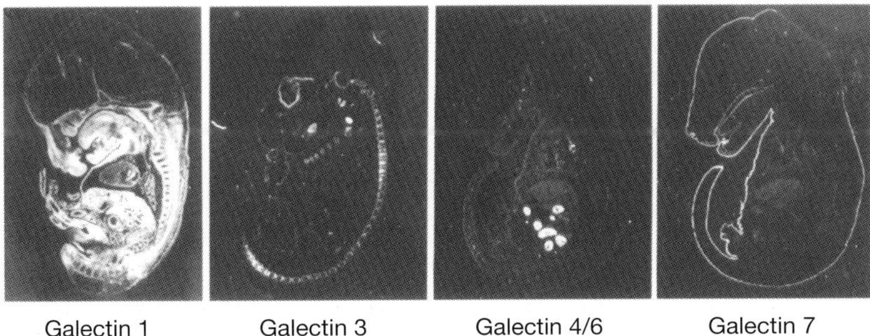

Figure 1 Expression patterns of galectins during mouse embryogenesis. Parasaggital sections of midgestation embryos were hybridized with ^{35}S-labelled RNA antisense probes.

In summary, a refined and dynamic repertoire of galectins is set up, starting early during mouse embryogenesis. In no case have the corresponding physiological glycosidic ligands been identified, but one can imagine that an equally refined and dynamic repertoire of glycosyltransferases is also established at the same time, presumably generating an immensely complex glycosidic 'map' of the embryo. Thus, a whole range of galectin–glycoside interactions might participate in many morphogenetic processes that take place during development. The identification of Fringe, a modulator of Notch, as a glycosyltransferase represents an example of how a specific glycosylation event may regulate developmental processes [22].

Galectins and embryonic implantation

We originally isolated galectin 1 in a differential screen that was designed to identify genes regulated at the time of embryonic implantation [23] and indeed the activation of galectin 1 expression in trophectoderm cells is concomitant with hatching of the blastocyst out of the zona pellucida (which is the ultimate step before attachment). This suggests a role for this lectin in the initial interactions between the embryo and the uterine wall [15]. Although the molecular events that mediate embryonic implantation remain to be elucidated, it has been well documented that carbohydrate residues located on the surface of the maternal tissue are involved in this key process of mammalian development [24]. However, when a null mutation was introduced into the mouse galectin 1 gene, it was found to be compatible with fertility and survival in animal house conditions [25]. Interestingly, the only other gene known to follow the same distinct activation pattern in E4.5 embryos is galectin 3. By pursuing our genetic approach, we found that a galectin 3 null mutation was also compatible with fertility and survival, and that even galectin 1/galectin 3 double mutants exhibit no overt phenotype [21]. This may not be too surprising since, as mentioned above, the two proteins have different subcellular localizations, even in the implanting embryo.

It remains formally possible that galectin 1 and/or galectin 3 normally participate in the process of implantation and that a backup mechanism operates in the mutants. This idea is supported indirectly by the fact that galectin 1 can accelerate the rate of blastocyst attachment on to endometrial cells *in vitro* (S. Kimber and F. Poirier, unpublished work). Alternatively, this crucial step of mammalian embryogenesis may be secured by several systems, all of which function normally and in parallel. If this is the case, the absence of one system (i.e. galectin–carbohydrate interactions) may have a very limited effect; for example, a short delay in the timing of implantation is conceivable.

Mutant phenotypes in *gal-1* $^{-/-}$ mice

The existence of mice that are homozygous galectin 1 mutants (*gal-1* $^{-/-}$ mice) have allowed us to start testing various hypotheses *in vivo* regarding galectin functions drawn from *in vitro* data.

A subtle, but robust, phenotype was discovered in the developing olfactory system. In the absence of galectin 1, a subpopulation of primary olfactory neurons fails to project to their correct target sites in the caudal olfactory bulb [26]. Thus, galectin 1, which is normally expressed transiently by ensheating cells of the olfactory nerve fibre at E15.5 [27], appears to be involved in the guidance, growth or maintenance of primary sensory olfactory axons between the nasal cavity and the olfactory bulb. This was the first evidence that a lectin has neurite outgrowth-promoting activity and plays a role in neuronal pathfinding in the mammalian nervous system. Potential functional consequences of this defect remain to be explored.

Galectin 1 had long been identified as an immunomodulating factor [28]. Most importantly, galectin 1 can induce apoptosis when added exogenously to human T-cells in culture [12]. Surprisingly, histological examination of the thymus and spleens from the *gal-1* $^{-/-}$ mice revealed no overt differences compared with wild-type mice. In the thymus of *gal-1* $^{-/-}$ mice, cortical and medullary regions were of similar size to those seen in the thymus from wild-type mice, with sharp corticomedullary junctions. In spleens from *gal-1* $^{-/-}$ mice, well-developed white pulp was present, and the number of lymphoid follicles did not differ from that observed in spleens of wild-type mice (L. Baum, personal communication).

Finally, it has been reported that galectin 1 is specifically induced in activated T-cells [29] and it is able to inhibit antigen-induced proliferation of naive and memory CD8+ T-cells *in vitro*. It was therefore proposed as an autocrine negative growth factor [29]; however, the galectin 1 null mutant mice were indistinguishable from wild-type littermates with respect to lymphocyte distribution and cytolytic T-lymphocyte activity after lymphocytic choriomeningitis virus infection (H. Pircher, personal communication).

Mutant phenotypes in *gal-3*$^{-/-}$ mice

Galectin 3 has also been implicated in the immune system, especially in inflammatory responses. Expression of galectin 3 has been described on neutrophils [30], eosinophils [31] and also on activated macrophages (Mac-2 marker) [17]. Therefore, as in the case of galectin 1, several studies focused on the immune system of galectin 3 mutants. The first set of experiments was to examine the response induced after triggering acute peritonitis by intraperitoneal injection of thioglycolate. We observed that recruitment of mutant granulocytes at the site of inflammation was normal, but that their maintenance inside the peritoneal cavity was impaired. Since neither massive apoptosis nor any major increase in macrophage phagocytic activity were detected, these results suggested the existence of an additional level of control during the resolution of acute inflammation [32]. It seems likely that galectin 3, which is normally present on the surface of granulocytes, participates in an active retention mechanism and prevents the cells from leaving the site of inflammation. Using similar experimental conditions, Hsu et al. [33] noticed an abnormal behaviour of galectin 3 mutant macrophages, suggesting another role for galectin 3 in macrophage cell shape, migration and possibly survival.

Several developmental aspects of the galectin 3 phenotype are under investigation, but the most extensive and informative study to date has been carried out in developing long bones because the galectin 3 gene is normally expressed from the earliest stage of bone formation (Figure 1). Adult galectin 3 mutants are of normal size and seem to have normal bones; however, there are many well-documented examples of major defects in bone development which get corrected at later stages. We therefore undertook a thorough analysis of the process of ossification in the absence of galectin 3.

We found that mutant long bones do indeed exhibit multiple, though subtle, defects during embryogenesis, both in the process of chondrogenesis and in the co-ordination between chondrogenesis and osteogenesis [34] (Figure 2). Several aspects of chondrogenesis are affected: abnormal glycogen aggregates appear in proliferating and maturating mutant chondrocytes, the synchrony of chondrocyte differentiation along the longitudinal axis of the bone is altered and, most importantly, the final stages of chondrocyte differentiation are accelerated as revealed by a size reduction of the terminal hypertrophic zone (Figures 3A–3D) and a higher incidence of 'dark' cells, i.e. cells with dense cytoplasm and nuclei containing finely condensed chromatin and large vacuoles. This latter point is particularly important because we have shown that these 'dark' cells correspond to the final step before chondrocyte death; however, these cells are not apoptotic, either by histological criteria (they contain many cytoplasmic organelles and no apoptotic bodies) or by the terminal transferase deoxytidyl uridine end-labelling assay. Taken together with recent results from a study of the rabbit growth plate [35], the study of the defects in long bones from *gal-3*$^{-/-}$ mice establishes that chondrocytes do not die by apoptosis. Instead, they follow an alternative pathway of programmed cell death, more similar to type II or lysosomal cell death as has been described during insect metamorphosis [36,37] or ovarian atrophy [38].

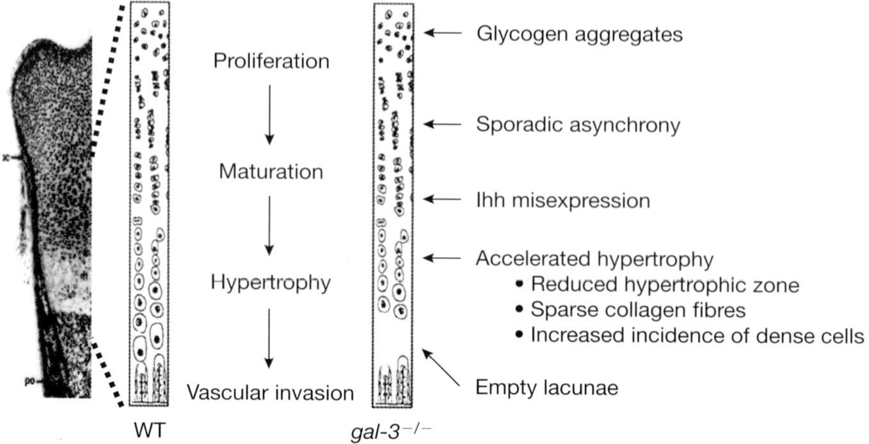

Figure 2 Schematic representation a growth plate from the femur of an E16.5 mouse embryo. The various defects of chondrogenesis observed in the galectin 3 mutants are indicated on the right-hand side. Ihh, Indian hedgehog. WT, wild-type; *gal-3*$^{-/-}$, galectin 3 homozygous mutant mice.

In addition, the various defects of chondrogenesis detected in the absence of galectin 3 also include limited and abnormal mineralization of the cartilage matrix, as well as sparse, short and irregular collagen fibres in the extracellular matrix, as revealed by electron microscopy (Figures 3E and 3F).

The most visible defect in the growth plate of long bones of *gal-3*$^{-/-}$ mice is the presence of empty lacunae at the junction between the last row of hypertrophic chondrocytes and the front of vascular invasion (Figures 3A–3D). This is a unique feature that is characteristic of *gal-3*$^{-/-}$ mutants and these results show, for the first time, that the processes of chondrogenesis and osteogenesis can be uncoupled. In other words, the triggering of programmed cell death in hypertrophic chondrocytes does not require direct contact with osteoclasts, osteoblasts or blood cells.

In conclusion, the panel of defects observed in growth plates from *gal-3*$^{-/-}$ mice provides new fundamental information that contributes to the understanding of the normal process of bone development. An important aspect of this study is the complexity of the mutant phenotype that, even within the bone, cannot *a priori* be explained by a single unique function of galectin 3. Rather, the range of defects implies that galectin 3 fulfils several functions during endochondral ossification, notably as a survival factor for hypertrophic chondrocytes and as a molecule that mediates, directly or indirectly, the coordination between chondrogenesis and osteogenesis. Although the precise molecular events underlying this phenotype remain to be elucidated, a systematic deregulation of Indian hedgehog expression in the bones of *gal-3*$^{-/-}$ mice has been documented, which suggests that the absence of galectin 3 somehow impinges on the control of this key regulator of chondrogenesis [39].

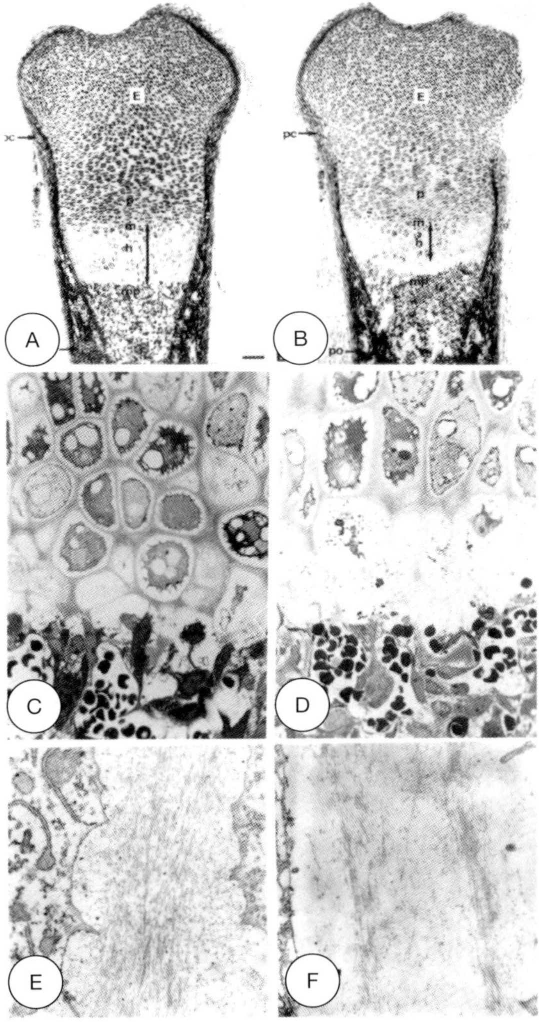

Figure 3 Histological comparison of wild-type and galectin 3 mutant embryonic femurs. Optical semi-thin sections (1 μm) from wild-type (**A** and **C**) and mutant (**B** and **D**) femurs were photographed under low (**A** and **B**) and high (**C** and **D**) magnification. Note the gap between the last row of hypertrophic chondrocytes and the vascular front in the mutant. Electron-microscopic sections from hypertrophic zones of wild-type and mutant mice are shown in (**E**) and (**F**) respectively. Zones of sparse and irregular collagen fibres are observed in the mutant.

Conclusions and perspectives

The mutational analysis of the galectin family is now shedding a different light on this class of molecules. On the one hand, it appears that galectins are not required for survival in animal-house conditions; on the other hand, there seem to be many subtle, but very complex, consequences of lacking galectins during development. A thorough description of the galectin null phenotypes may require extensive examination, including electron microscopy studies. For instance, just as bone development is abnormal in galectin 3 mutants, even though the adult bones appear normal, so too the process of embryonic implantation might be abnormal in the absence of galectin 1 and/or galectin 3, even though the implantation succeeds in the end. Several interpretations may explain a large panel of mild defects. A possibility remains that strict redundancy is taking place between different members of the galectin family, although we do not have any indication supporting this hypothesis. In this case, the prediction would be that compound galectin mutants would exhibit more defects than the sum of individual mutants. Another, perhaps more attractive, possibility is that parallel systems operate in many developmental processes so that the lack of a galectin can be compensated for by the deregulation of one or several parallel systems. Such separate parallel systems, which are yet to be identified, would presumably be different depending on the biological process (i.e. implantation, olfaction or ossification).

At this stage of the analysis, the emerging picture is that in normal conditions galectins function as 'optimizing' molecules, i.e. once removed, the system still works only slightly less efficiently. This has no major consequences for the individual, but it may have some for the population in the long run. In addition, it is now important to evaluate the response of galectin mutants to stress or pathological conditions; for example, the lack of galectin 3 in developing bones might become detrimental for the animal in the case of a vitamin-D-deficient diet, or during the repair mechanism following a fracture.

In view of the complexity of galectin phenotypes, the precise definition of the cell type, subcellular compartment and stage of differentiation when a galectin operates *in vivo* will be an important step towards isolating their molecular partners in the cell.

I thank all past and present members of the laboratory. This work is supported by the CNRS (Centre National de la Recherche Scientifique) and by the ARC (Association de la Recherche sur le Cancer).

References

1. Barondes, S.H., Cooper, D.N.W., Gitt, M.A. and Leffler, H. (1994) J. Biol. Chem. **269**, 20807–20810
2. Barondes, S.H., Castronovo, V., Cooper, D.N., Cummings, R.D., Drickamer, K., Feizi, T., Gitt, M.A., Hirabayashi, J., Hughes, C., Kasai, K. et al. (1994) Cell **76**, 597–598
3. Drickamer, K. (1988) J. Biol. Chem. **263**, 9557–9560
4. Cooper, D.N. and Barondes, S.H. (1999) Glycobiology **9**, 979–984
5. Hughes, R.C. (1999) Biochim. Biophys. Acta **1473**, 172–185
6. Perillo, N.L., Marcus, M.E. and Baum, L.G. (1998) J. Mol. Med. **76**, 402–412
7. Leffler, H. (2001) Results Probl. Cell Differ. **33**, 57–83
8. Cooper, D.N., Massa, S.M. and Barondes, S.H. (1991) J. Cell Biol. **115**, 1437–1448
9. Sato, S. and Hughes, R.C. (1992) J. Biol. Chem. **267**, 6983–6990
10. Zhou, Q. and Cummings, R.D. (1993) Arch. Biochem. Biophys. **300**, 6–17
11. Kuwabara, I. and Liu, F.T. (1996) J. Immunol. **156**, 3939–3944

12. Perillo, N.L., Pace, K.E., Seilhamer, J.J. and Baum, L.G. (1995) Nature (London) **378**, 736–739
13. Akahani, S., Nangia-Makker, P., Inohara, H., Kim, H.R. and Raz, A. (1997) Cancer Res. **57**, 5272–5276
14. Dagher, S.F., Wang, J.L. and Patterson, R.J. (1995) Proc. Natl. Acad. Sci. U.S.A. **92**, 1213–1217
15. Poirier, F., Timmons, P.M., Chan, C.T., Guenet, J.L. and Rigby, P.W. (1992) Development **115**, 143–155
16. Fowlis, D., Colnot, C., Ripoche, M.A. and Poirier, F. (1995) Dev. Dyn. **203**, 241–251
17. Ho, M.K. and Springer, T.A. (1982) J. Immunol. **128**, 1221–1228
18. Gitt, M.A., Colnot, C., Poirier, F., Nani, K.J., Barondes, S.H. and Leffler, H. (1998) J. Biol. Chem. **273**, 2954–2960
19. Timmons, P.M., Colnot, C., Cail, I., Poirier, F. and Magnaldo, T. (1999) Int. J. Dev. Biol. **43**, 229–235
20. Gitt, M.A., Xia, Y.R,. Atchison, R.E., Lusis, A.J., Barondes, S.H. and Leffler, H. (1998) J. Biol. Chem. **273**, 2961–2970
21. Colnot, C., Fowlis, D., Ripoche, M.A., Bouchaert, I. and Poirier, F. (1998) Dev. Dyn. **211**, 306–313
22. Moloney, D.J., Panin, V.M., Johnston, S.H., Chen, J., Shao, L., Wilson, R., Wang, Y., Stanley, P., Irvine, K.D., Haltiwanger, R.S. and Vogt, T.F. (2000) Nature (London) **406**, 369–375
23. Poirier, F., Chan. C.T., Timmons, P.M., Robertson, E.J., Evans, M.J. and Rigby, P.W. (1991) Development **113**, 1105–1114
24. Poirier, F. and Kimber, S. (1997) Mol. Hum. Reprod. **3**, 907–918
25. Poirier, F. and Robertson, E.J. (1993) Development **119**, 1229–1236
26. Puche, A.C., Poirier, F., Hair, M., Bartlett, P.F. and Key, B. (1996) Dev. Biol. **179**, 274–287
27. Tenne-Brown, J., Puche, A.C. and Key, B. (1998) Int. J. Dev. Biol. **42**, 791–799
28. Levi, G., Tarrab-Hazdai, R. and Teichberg, V.I. (1983) Eur. J. Immunol. **13**, 500–507
29. Blaser, C., Kaufmann, M., Muller, C., Zimmermann, C., Wells, V., Mallucci, L. and Pircher, H. (1998) Eur. J. Immunol. **28**, 2311–2319
30. Truong, M.J., Gruart. V., Kusnierz, J.P., Papin, J.P., Loiseau, S., Capron, A. and Capron, M. (1993) J. Exp. Med. **177**, 243–248
31. Truong, M.J., Gruart, V., Liu, F.T., Prin, L., Capron, A. and Capron, M. (1993) Eur. J. Immunol. **23**, 3230–3235
32. Colnot, C., Ripoche, M.A., Milon, G., Montagutelli, X., Crocker, P.R. and Poirier, F. (1998) Immunology **94**, 290–296
33. Hsu, D.K., Yang, R.Y., Pan, Z., Yu, L., Salomon, D.R., Fung-Leung, W.P. and Liu, F.T. (2000) Am. J. Pathol. **156**, 1073–1083
34. Colnot, C., Sidhu, S.S., Balmain, N. and Poirier, F. (2001) Dev. Biol. **229**, 203–214
35. Roach, H.I. and Clarke, N.M. (2000) J. Bone Joint Surg. Br. **82**, 601–613
36. Bowen, I.D., Mullarkey, K. and Morgan, S.M. (1996) Microsc. Res. Techniques **34**, 202–217
37. Lockshin, R.A. and Zakeri, Z. (1994) Biochem. Cell. Biol. **72**, 589–596
38. D'Herde, K., De Prest, B. and Roels, F. (1996) Reprod. Nutr. Dev. **36**, 175–189
39. St-Jacques, B., Hammerschmidt, M. and McMahon, A.P. (1999) Genes Dev. **13**, 2072–2086

Biochem. Soc. Symp. **69**, 105–115
(Printed in Great Britain)

9

MS screening strategies: investigating the glycomes of knockout and myodystrophic mice and leukodystrophic human brains

Mark Sutton-Smith*, Howard R. Morris*, Prabhjit K. Grewal†, Jane E. Hewitt†, Reginald E. Bittner‡, Ehud Goldin§, Raphael Schiffmann§ and Anne Dell*[1]

*Department of Biological Sciences, Imperial College of Science, Technology and Medicine, London SW7 2AY, U.K., †Institute of Genetics, Queen's Medical Centre, University of Nottingham, Nottingham NG7 2UH, U.K., ‡Neuromuscular Research Department, Institute of Anatomy, University of Vienna, Waehringerstrasse 13, 1090 Vienna, Austria, and §National Institutes of Health, Building 10, 9000 Rockville Pike, Bethesda, MD 20892-1260, U.S.A.

Abstract

The implementation of highly sensitive and rapid mass spectrometric screening strategies for defining the glycosylation repertoires of organs in knockout mice is helping to reveal the roles that glycans play in health and disease. Thus novel glycosylation pathways have been uncovered in two such knockouts, namely α-mannosidase II null mice and UDP-N-acetylglucosamine:α6-D-mannoside β1,2-N-acetylglucosaminyltransferase II null mice. This chapter documents the glycosylation profiles of a wide range of organs from the normal mouse which should facilitate future glycomics studies of knockout mice. Furthermore, we report applications of our screening technology in studies of the myodystrophy mouse and a human leukodystrophy.

[1]To whom correspondence should be addressed (e-mail a.dell@ic.ac.uk).

Introduction

In the post-genome era, attention is being focused on the post-translational events that yield functionally active gene products. Among these, glycosylation is by far the most abundant, and the majority of proteins both inside and outside the cell are glycosylated. Glycan moieties are implicated in a wide range of intracellular, cell–cell and cell–matrix recognition events [1,2]. Although much is known about the structures of glycosylated polymers in a very wide range of organisms (see, for example, the structural data provided at the following websites: www.glycosuite.com and www.glyco.org), and many structural tools are available for probing glycan structure, there is a need for methodology which will rapidly provide information on the glycan repertoire (the glycome) of cells, tissues, organs etc. A variety of strategies for mapping glycomes are currently being pursued by different laboratories. These include lectin- [3] and HPLC-profiling [4,5], and nano-NMR analysis of glycan mixtures [4]. In this chapter, we describe MS screening strategies that we have developed for rapid and sensitive determination of the glycan content of biological materials [6]. Our strategies are illustrated with data from three lines of research: (i) studies of knockout mice; (ii) analysis of the myodystrophic (*myd*) mouse; and (iii) searching for glycosylation disorders in a human leukodystrophy.

MS screening strategies

In recent years, we have developed rapid, high-sensitivity MS strategies for screening for the types of glycans present in a diverse range of biological material, including body fluids, secretions, organs, cultured cell lines and whole parasites [6–8]. These methods are based on the analysis of permethylated derivatives, which yield molecular ions at high sensitivity, irrespective of the type of ionization [fast atom bombardment (FAB), matrix-assisted laser desorption or electrospray], and which, in FAB–MS experiments, reliably afford characteristic fragment ions (called A-type ions) resulting from cleavage at *N*-acetylhexosamine residues. The 'map' of A-type ions generated from a mixture of glycans defines all the non-reducing sequences present in the sample. Putative structures are assigned to each molecular ion based on the usually unique glycan composition for a given mass, the non-reducing epitopes defined by the A-type fragment ions and prior knowledge of N- and O-glycan biosynthesis. Assignments can then be confirmed by MS experiments on chemical and enzymic degradations, the choice of which is guided by the sequence information provided by the screening experiments.

Structural glycobiology of the normal and knockout mouse

Investigations of knockout mice are revealing important aspects of glycan functions in mammalian systems [9,10]; however, mouse phenotypes are often difficult to interpret because the precise nature of glycan alterations resulting from ablation of genes encoding for biosynthetic enzymes is unknown. In addition, although the glycosylation of some murine glycoproteins is well established, the glycan repertoires of many mouse organs are poorly defined. In a previous report, we have documented the N- and O-glycosylation profiles revealed by MS screening of the brain, kidney and lung of normal mice [6] and in studies of two knockout mice we have shown that novel glycosylation pathways can result from gene ablation [11,12]. Thus MS screening of α-mannosidase II null mice revealed the presence in the kidney of novel hybrid glycans with fucosylated, bisected cores carrying Lewis^{x} antennae [11]. Notably, the knockout mice exhibit an autoimmune disease that is similar to human systemic lupus erythematosus and it has been observed that anti-self antibodies with reactivity towards the novel N-glycans are induced in these mice [11]. This is an important discovery because it suggests that abnormal glycosylation could be a cause of some systemic autoimmune diseases. Novel structures have also been found in the kidney of the UDP-N-acetylglucosamine:α6-D-mannoside β1,2-N-acetylglucosaminyltransferase II null mouse [12]. This enzyme is responsible for initiating the second antenna of complex-type glycans and the knockout is a model for one of the congenital disorders of glycosylation in humans: congenital disorders of glycosylation type IIa [13]. As expected, the mouse lacks glycans with antennae on the upper mannose of the core. Unexpectedly, however, our structural studies [12] have revealed that the mouse kidney is rich in 'pseudo-biantennary' glycans resulting from elongation of the bisecting *N*-acetylglucosamine (GlcNAc) residue. This is the first documented example of elongation of the bisecting GlcNAc of an N-glycan.

Documenting the glycome of the normal mouse

As indicated above, we have previously reported glycosylation profiles obtained in screening experiments on the murine brain, kidney and lung. To complement this information, Figure 1 shows data derived from the liver, spleen, small intestine, colon, stomach, pancreas, lung, thymus, ovaries and testis of the normal mouse.

Analysis of the *myd* mouse

Several lines of recent research have arrived at the important conclusion that glycosylation abnormalities may play a pivotal role in congenital muscular dystrophies. The first evidence linking aberrant glycosylation to muscular dystrophy came from studies of the myd mouse [14]. The *myd* mouse results from a spontaneous mutation that produces an autosomal recessive neuromuscular phenotype. Recently, the *myd* mutation was identified as a genomic deletion of >100 kb within the Large gene, which is believed to encode a glycosyltransferase [15]. This deletion removes exons 5–7, causing a frameshift in the resulting mRNA and premature termination of the protein. Examination of the genomic sequence gave no indication of any other gene residing within the deleted sequence; therefore, loss of functional Large protein appears to be the causative mutation in *myd*. Unusually, Large has two putative catalytic regions which each

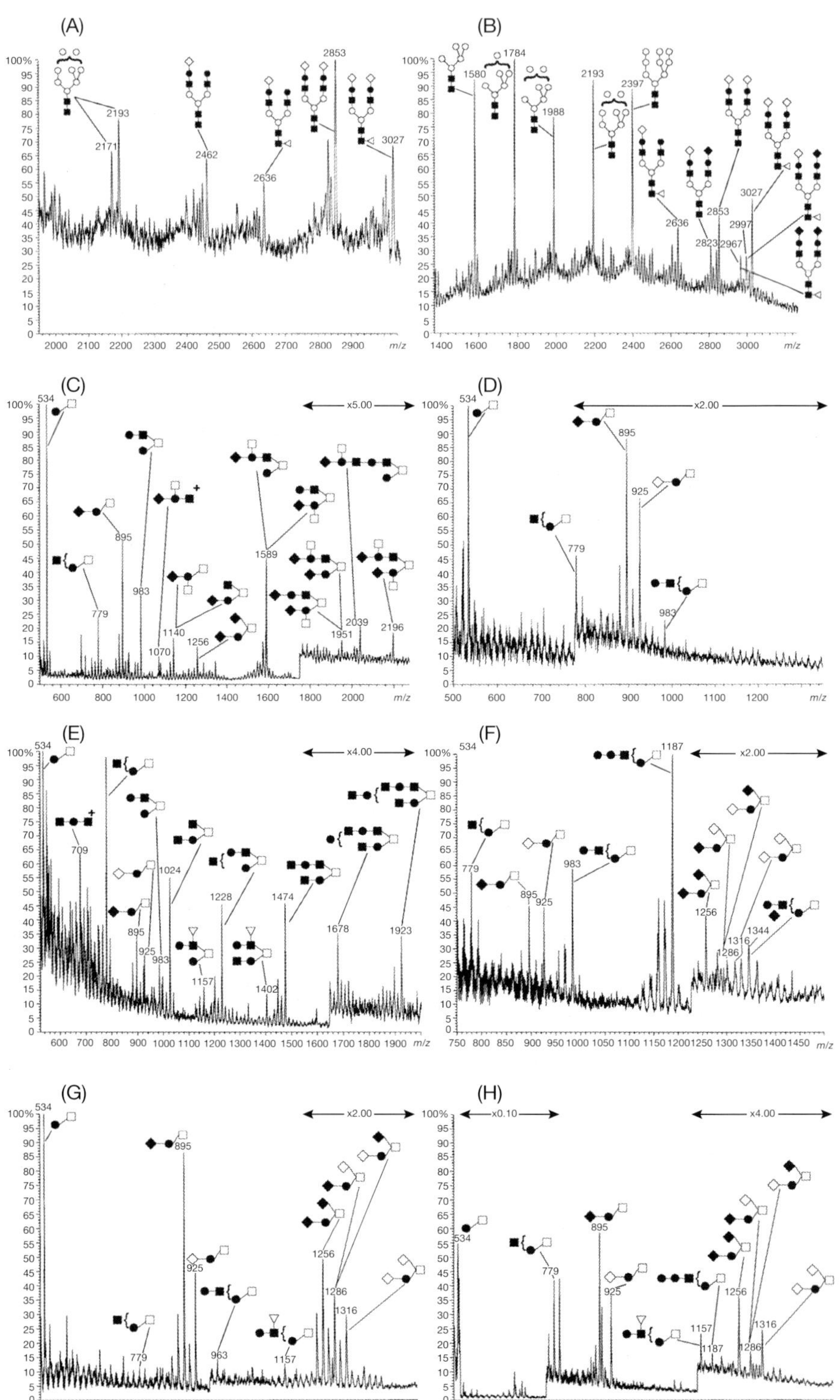

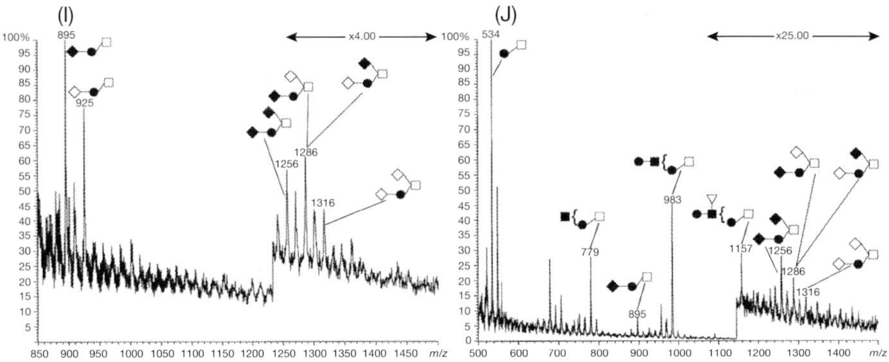

Figure 1 FAB mass spectra of permethylated glycans from various normal mouse organs. FAB mass spectra of molecular ion region of permethylated N-glycans from normal mouse liver (**A**) and spleen (**B**). FAB mass spectra of permethylated O-glycans from normal mouse small intestine (**C**), colon (**D**), stomach (**E**), pancreas (**F**), lung (**G**), thymus (**H**), ovaries (**I**) and testes (**J**). ●, Galactose; □, GalNAc; ■, GlcNAc; △, fucose; ◆, NeuAc; ◇, NeuGc.

contain the Asp-Xaa-Asp motif that is typical of glycosyltransferases that use nucleoside diphosphate sugars as donors [16]. The presence of two catalytic domains within a single glycosyltransferase enzyme is unusual, although an increasing number of such enzymes are being identified, including class 2 hyaluronate synthase [17] and a cytosolic α1,2-fucosyltransferase from *Dictyostelium* [18]. As the biochemical activity of Large has not yet been determined, homology with known glycosyltransferases is the only way to infer what donors and acceptors it might use. Catalytic domain 1 is most closely related (approx. 22% amino acid identity) to the bacterial WaaIJ family of putative α-glycosyltransferases [19]. Members of this family are involved in synthesis of the bacterial outer membrane lipopolysaccharide. Where substrates for these proteins are known, they use UDP-galactose or UDP-glucose. Catalytic domain 2, in contrast, is most closely related to β-1,3-N-acetylglucosaminyltransferase, with 28% amino acid identity. This enzyme is required for synthesis of the poly(*N*-acetyl lactosamine) backbone (Galβ1–4GlcNAcβ1–3)$_n$, part of the erythrocyte i-antigen. The presence of two domains within the same protein suggests Large might be involved in addition of a disaccharide unit.

Western blotting, using a monoclonal antibody that recognizes a carbohydrate epitope on mouse α-dystroglycan (α-DG), has provided evidence that the glycosylation of α-DG is altered in *myd* in all tissues analysed, including brain [14]. α-DG is a key component of the dystrophin-associated glycoprotein complex [20]. It is synthesized as part of a precursor polypeptide that is post-translationally cleaved and differentially glycosylated to yield α- and β-DGs. The latter is a transmembrane glycoprotein that binds dystrophin and the intracellular cytoskeleton. In the basal lamina of muscle cells and Schwann cells, α-DG binds both β-DG and extracellular laminin, and acts as a link between them [21]. α-DG is a heavily glycosylated protein carrying both N- and O-glycans and possibly glycosaminoglycan modifications. The α-DG–laminin interaction is believed to involve a sialylated mucin-like domain in the centre of α-DG that, in bovine peripheral nerve and rabbit skeletal muscle, has been shown to carry a cluster of novel O-mannosyl-linked sialylated tetrasaccharides (NeuAcα2–3Galβ1–4GlcNAcβ1–2Man) [22]. Notably, the human GlcNAc transferase that catalyses the addition of GlcNAc to the O-linked mannose has recently been cloned and expressed [23]. Importantly, six patients with muscle–eye–brain disease (MEB), a disease characterized by congenital muscular dystrophy, are now known to have point mutations in this GlcNAc transferase gene [24]. Thus, aberrant glycosylation of α-DG is likely to play a role in MEB and, indeed, in several other congenital muscular dystrophies where abnormal α-DG has been detected by Western blotting [25,26].

Whether or not α-DG from the *myd* mouse carries aberrant O-mannosyl glycans remains to be established. Indeed, α-DG glycosylation in the normal mouse has yet to be defined. The determination of glycan structures that are abnormal or missing in *myd* would help to define the biochemical activity of Large, as well as identifying functionally important carbohydrate structures on α-DG. Based on previous structural studies of bovine, rabbit and porcine α-DG [22,27], it is probable that both mannose- and *N*-acetylgalactosamine (GalNAc)-linked O-linked glycans are present in murine α-DG.

We are currently engaged in isolating α-DG from the *myd* mouse for structural characterization. This is a challenging long-term programme of work because only tiny amounts of material can be obtained from each mouse. Therefore, we are also employing our screening strategies to search for global glycosylation abnormalities because, as mentioned earlier, these methods provide information very rapidly. The respective N- and O-glycan profiles of the kidney and brain are shown in Figure 2. The molecular ion regions spanning the mass range corresponding to the major bi-, tri- and tetra-antennary N-glycans of the kidney are shown in Figure 2(A) (normal mouse) and Figure 2(B) (*myd* mouse). All the signals observed in the normal mouse are present at similar abundance in the *myd* mouse and there is no evidence for novel glycosylation. These results suggest that antenna-initiating GlcNAc transferases, as well as the bisecting GlcNAc transferase, are not altered in the *myd* kidney. The normal mouse brain is rich in O-mannosyl-linked oligosaccharides, in addition to normal GalNAc-linked O-glycans, and is therefore a good choice of organ for preliminary screening for changes in O-mannosyl glycans. The major O-mannosyl-linked glycan that was detected in the normal mouse brain (*m/z* 912 in Figure 2C) corresponds to the Lewis^{x}-containing glycan that had been identified previously in α-DG from sheep brain. Also present, but at lower abundance, is the sialylated O-mannosyl-linked glycan (*m/z* 1099 in Figure 2C). Signals corresponding to GalNAc-linked oligosaccharides are present at *m/z* 895, 1157 and 1256. The *myd* mouse brain afforded the same molecular ions as the normal (compare Figure 2C with Figure 2D), but there is a small reduction in the abundance of *m/z* 912 when compared with GalNAc-terminated glycans of a similar mass. This is an intriguing observation because it could suggest that the Large mutation is affecting some O-mannosyl-linked

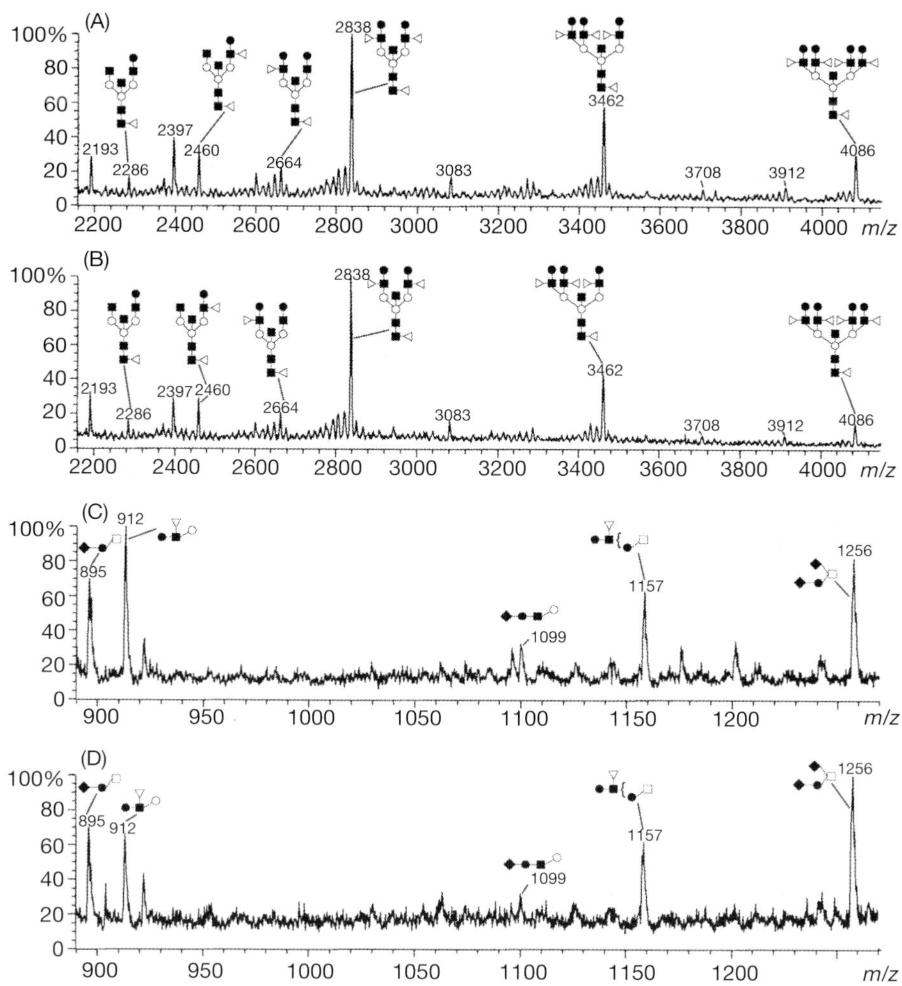

Figure 2 N- and O-glycan profiles of normal and *myd* mouse kidney and brain, FAB mass spectra of molecular ion region of permethylated N-glycans from normal (**A**) and *myd* (**B**) mouse kidney. FAB mass spectra of permethylated O-glycans from normal (**C**) and *myd* (**D**) mouse brain. ●, Galactose; ○, mannose; □, GalNAc; ■, GlcNAc; △, fucose; ◆, NeuAc.

glycosylation; however, further experiments are required to establish the reproducibility of this apparent reduction in abundance.

The kidney- and brain-screening data provide unequivocal evidence for the integrity of key N- and O-glycosylation pathways in the *myd* mouse. These include the biosynthesis of complex-type glycans with lactosamine antennae and small O-glycans attached to proteins via mannose or GalNAc. Experiments are underway to establish whether polylactosamine biosynthesis in N- and/or O-glycans has been affected by the *myd* mutation. Although it is clear from the brain screen that bulk O-mannosyl glycans are unaffected in the *myd* mouse, it remains a possibility that Large specifically glycosylates α-DG and that its O-mannosyl glycans are abnormal. However, in view of the human MEB disease data described earlier, in which the O-mannosyl-associated GlcNAc transferase has been identified, it is more likely that Large is associated with other types of glycosylation in α-DG, perhaps polylactosamine or glycosaminoglycan biosynthesis. Structural studies to explore these possibilities are underway in our laboratories.

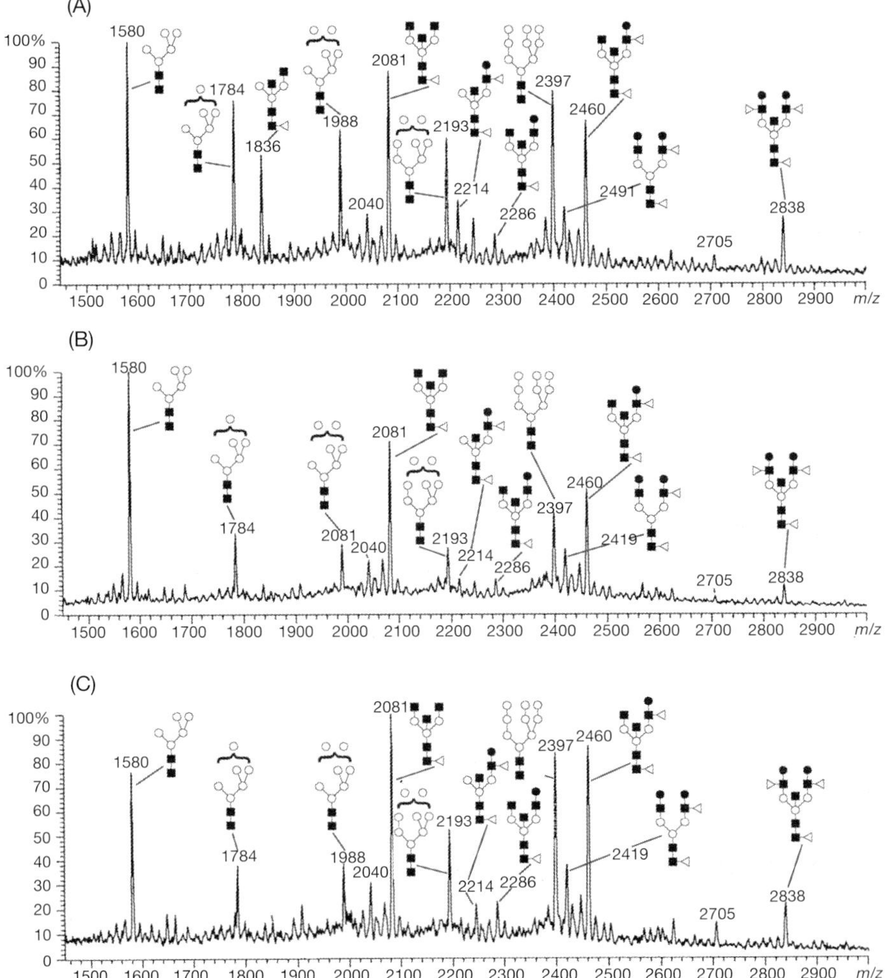

Figure 3 N-glycan profiles of normal mouse brain and the brains of two CACH patients. FAB mass spectra of molecular ion region of permethylated N-glycans from normal mouse brain (**A**) and the brains of CACH patients 1 (**B**) and 2 (**C**). ●, Galactose; ○, mannose; ■, GlcNAc; △, fucose.

Childhood ataxia with central nervous system hypomyelination (CACH)

CACH, also termed 'leukodystrophy with vanishing white matter', is an autosomal genetic disorder of brain white matter (leukodystrophy) that occurs mainly in children [28]. Until very recently, nothing was known about the aetiology of this disease; however new research [29] has found that mutations in the EIF2B5 and EIF2B2 genes, encoding the ϵ- and β-subunits respectively of the translation initiation factor eIF2B, occur in CACH patients. eIF2B is a guanine nucleotide-exchange protein that consists of five subunits, which functions in

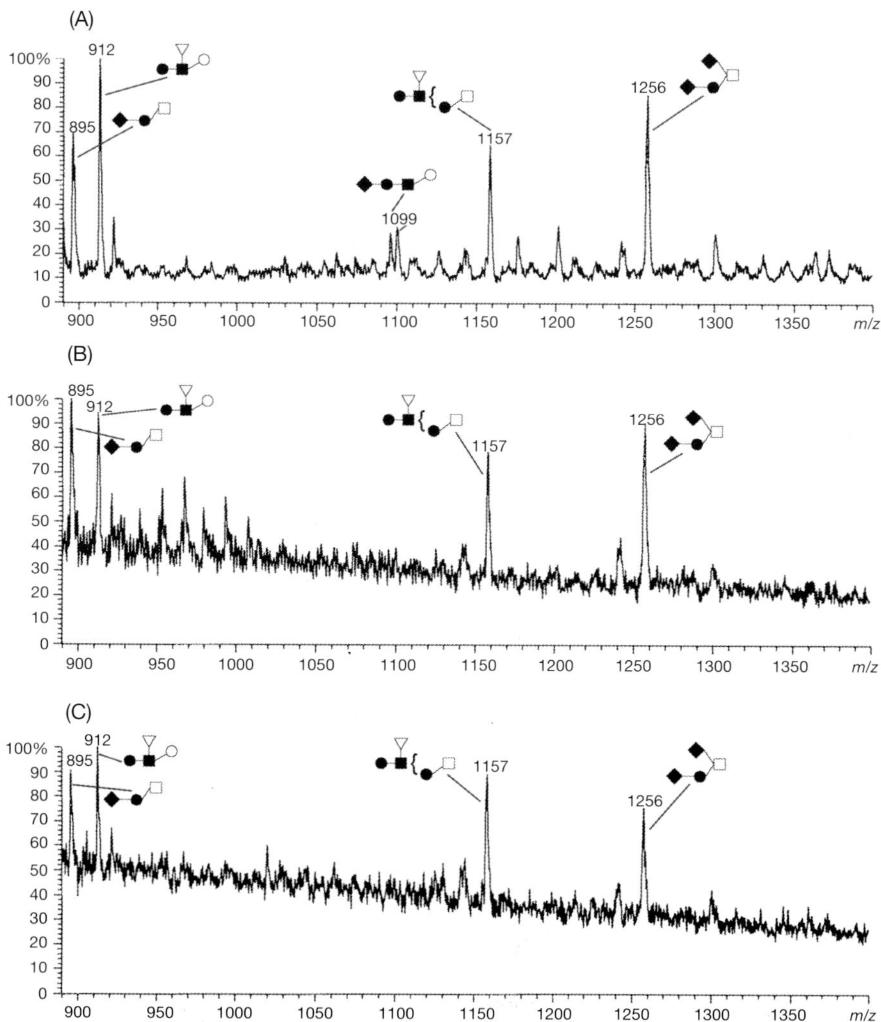

Figure 4 O-glycan profiles of normal mouse brain and the brains of two CACH patients. FAB mass spectra of permethylated O-glycans from normal mouse brain (**A**) and the brains of CACH patients 1 (**B**) and 2 (**C**). ●, Galactose; ○, mannose; ■, GlcNAc; △, fucose; ◆, NeuAc.

the regulation of translation in eukaryotic cells. The majority of the CACH mutations are associated with the ε-subunit which can recapitulate the catalytic function of intact eIF2B *in vitro*. Because of the pivotal role that eIF2B plays in the regulation of translation, it is not surprising that mutations in the genes for this protein have devastating effects on affected individuals

The neuropathology of CACH is characterized by the presence of foamy oligodendroglial-like cells and distorted astrocytes. We have found that the cytoplasm of these foamy cells show abnormal staining with Alcian Blue, a dye that is known to bind to anionic polysaccharides. With the aim of providing a biochemical explanation for this observation, we have initiated a structural study of brain glycopolymers in CACH patients in order to determine whether altered glycosylation occurs in this disease. Results from our screening of N- and O-glycans are presented below.

Figures 3 and 4 show the molecular ion regions of spectra obtained from mixtures of N- and O-glycans, respectively, isolated from the brains of two CACH patients. Data from normal mouse brains obtained using the same screening protocols are shown in the upper panel of each figure. The mouse was used as a 'control' in these experiments because previous structural studies have suggested that rodent and human brains exhibit similar glycosylation profiles [30]. The data from the brains of both CACH patients are in accord with a normal profile, taking into account published structural data [30] and the results of our mouse screening. Thus, we conclude that the biosynthetic enzymes involved in processing the N-glycan precursor, initiating antennae formation and fucosylating the core and antennae of complex-type N-glycans are normal in CACH. Similarly, enzymes involved in O-glycan biosynthesis in the brain appear to be normal. Importantly, Figure 4 provides clear evidence for the presence of O-mannosyl structures in the human brain. As noted in the previous section, the existence of this family of glycans in the human has been inferred from the research on congenital muscular dystrophy, but no human O-mannosyl structures have been previously characterized. Our results suggest that, as in the pig [27], Lewis^{x}-containing structures are the main O-mannosyl glycans in mouse and human brains.

This work was supported by the Biotechnology and Biological Sciences Research Council and by the Wellcome Trust.

References

1. Varki, A. (1993) Glycobiology **3**, 97–130
2. Wells, L., Vosseller, K. and Hart, G.W. (2001) Science **291**, 2376–2378
3. Hirabayashi J., Arata Y. and Kasai, K. (2001) Proteomics **1**, 295–303
4. Manzi A.E., Norgard-Sumnicht K., Argade S., Marth J.D., van Halbeek H. and Varki A. (2000) Glycobiology. **10**, 669–689
5. Norgard-Sumnicht, K., Bai, X., Esko, J.D., Varki, A. and Manzi, A.E. (2000) Glycobiology **10**, 691–700
6. Sutton-Smith, M., Morris, H.R. and Dell, A. (2000) Tetrahedron Assymetry **11**, 363–369
7. Dell, A. and Morris, H.R.(2001) Science **291** 2351–2356
8. Haslam, S.M., Morris, H.R. and Dell, A. (2001) Trends Parasitol. **17**, 231–235

9. Marth, J.D. (1994) Glycoconjugate J. **1**, 3–8
10. Stanley, P. and Ioffe, E. (1995) FASEB J. **9**, 1436–1444
11. Chui, D., Sellakumar, G., Green, R.S., Marek, K.W., Sutton-Smith, M., McQuistan T., Morris, H.R., Dell, A. and Marth, J.D. (2001) Proc. Natl. Acad. Sci. U.S.A. **98**, 1142–1147
12. Wang, Y., Tan, J., Sutton-Smith, M., Ditto, D., Panico, M., Campbell, R.M., Varki, N.M., Long, J.M., Jaeken, J., Levinson, S.R. et al. (2001) Glycobiology **11**, 1051–1070
13. Schachter, H. and Jaeken, J. (1999) Biochim. Biophys. Acta **1455**, 179–192
14. Grewal, P.K., Holzfeind, P.J., Bittner, R.E. and Hewitt, J.E. (2001) Nat. Genet. **28**, 151–154
15. Peyrard, M., Seroussi, E., Sandberg-Nordqvist, A.-C., Xie, Y.-G., Han, F.-Y., Fransson, I., Collins, J., Dunham, I., Kost-Alimova, M. and Imreh, S. (1999) Proc. Natl. Acad. Sci. U.S.A. **96**, 598–603
16. Breton, C. and Imberty, A. (1999) Curr. Opin. Struct. Biol. **9**, 563–571
17. Jing, W. and DeAngelis, P.L. (2000) Glycobiology **10**, 883–889
18. van der Wel, H., Morris, H.R., Panico, M., Paxton, T., North, S.J., Dell, A., Thomson, J.M. and West C.M. (2001) J. Biol. Chem. **276**, 33952–33963
19. Heinrichs, D.E., Yethon, J.A. and Whitfield, C. (1998) Mol. Microbiol. **30**, 221–232
20. Durbeej, M., Henry, M.D. and Campbell, K.P. (1998) Curr. Opin. Cell Biol. **10**, 594–601
21. Ervasti, J.M. and Campbell, K.P. (1993) J. Cell Biol. **122**, 809–823
22. Endo, T. (1999) Biochim. Biophys. Acta **1473**, 237–246
23. Zhang, W., Betel, D. and Schachter, H. (2002) Biochem. J. **361**, 153–162
24. Yoshida, A., Kobayashi, K., Manya, H., Taniguchi, K., Kano, H., Mizuno, M., Inazu, T., Mitsuhashi, H., Takahashi, S., Takeuchi, M. et al. (2001) Dev. Cell. **1**, 717–724
25. Brockington, M., Blake, D.J., Prandini, P., Brown, S.C., Torelli, S., Benson, M.A., Ponting, C.P., Estournet, B., Romero, N.B., Mercuri, E. et al. (2001) Am. J. Hum. Genet. **69**, 1198–1209
26. Brockington, M., Yuva, Y., Prandini, P., Brown, S.C., Torelli, S., Benson, M.A., Herrmann, R., Anderson, L.V., Bashir, R., Burgunder, J.M. et al. (2001) Hum. Mol. Genet. **10**, 2851–2859
27. Smalheiser, N.R., Haslam, S.M., Sutton-Smith, M., Morris, H.R. and Dell, A. (1998) J. Biol. Chem. **273**, 23698–23703
28. Schiffmann, R. and Boespflug-Tanguy, O. (2001) Curr. Opin. Neurol. **14**, 789–794
29. Leegwater, P.A., Vermeulen, G., Konst, A.A., Naidu, S., Mulders, J., Visser, A., Kersbergen, P., Mobach, D., Fonds, D., van Berkel, C.G. et al. (2001) Nat. Genet. **29**, 383–388
30. Albach, C., Klein, R.A. and Schmitz, B. (2001) Biol. Chem. **382**, 187–194

Biochem. Soc. Symp. **69**, 117–134
(Printed in Great Britain)

10

The glycomes of *Caenorhabditis elegans* and other model organisms

Stuart M. Haslam*[1], David Gems†, Howard R. Morris* and Anne Dell*

*Department of Biological Sciences, Imperial College of Science, Technology and Medicine, London SW7 2AY, U.K., and †Department of Biology, University College London, London NW1 2HE, U.K.

Abstract

There is no doubt that the immense amount of information that is being generated by the initial sequencing and secondary interrogation of various genomes will change the face of glycobiological research. However, a major area of concern is that detailed structural knowledge of the ultimate products of genes that are identified as being involved in glycoconjugate biosynthesis is still limited. This is illustrated clearly by the nematode worm *Caenorhabditis elegans*, which was the first multicellular organism to have its entire genome sequenced. To date, only limited structural data on the glycosylated molecules of this organism have been reported. Our laboratory is addressing this problem by performing detailed MS structural characterization of the N-linked glycans of *C. elegans*; high-mannose structures dominate, with only minor amounts of complex-type structures. Novel, highly fucosylated truncated structures are also present which are difucosylated on the proximal *N*-acetylglucosamine of the chitobiose core as well as containing unusual Fucα1–2Gal1–2Man as peripheral structures. The implications of these results in terms of the identification of ligands for genomically predicted lectins and potential glycosyltransferases are discussed in this chapter. Current knowledge on the glycomes of other model organisms such as *Dictyostelium discoideum*, *Saccharomyces cerevisiae* and *Drosophila melanogaster* is also discussed briefly.

[1]To whom correspondence should be addressed (e-mail s.haslam@ic.ac.uk).

Introduction

There is now irrefutable evidence that glycoconjugates are important contributors to a huge range of cell–cell, cell–matrix and even intracellular events. As we attempt to better comprehend these events, a knowledge of the structures of the participants is vital. The recent increase in the number of genome and proteome sequencing projects will provide large amounts of information about the proteins that are present. However, a complete understanding of the role of glyconjugates in an organism will require the deciphering of its glycome, i.e. the entire set of glycans that are present in the organism. It is likely that the first organisms for which glycomes will be defined will be the model organisms that are already used commonly by biologists, e.g. *Caenorhabditis elegans*, *Dictyostelium discoideum*, *Saccharomyces cerevisiae* and *Drosophila melanogaster*, especially as the genome sequences of all these organisms have been or will very shortly be completely determined.

This chapter reviews the current knowledge of the glycome of these model organisms, with particular emphasis on *C. elegans*.

C. elegans

The nematode worm *C. elegans* is a small, multicellular, transparent, rapidly growing organism. It was the first multicellular organism to have its entire genome sequenced [1] and has had its entire cell lineage mapped [2]. It is therefore a valuable experimental model organism in developmental and molecular cell biology [3].

In recent years, there has been an increasing interest in the glycobiology of *C. elegans* and it is our belief that, of all the model organisms discussed in this review, *C. elegans* will have an impact on glycobiology that will be second only to that of the mouse. A link between glycosylation and *C. elegans* morphological development has been demonstrated. Three genes, *sqv-3, sqv-7* and *sqv-8*, which are required for invagination of vulval epithelial cells, oocyte formation and embryogenesis, encode enzymes with galactosyl transferase I, nucleotide-sugar transporter and glucuronyl transferase I activity respectively. All three enzymes are required for glycosaminoglycan biosynthesis [4,5]. The sequencing of the *sqv-3* gene also allowed the identification of a human homologue which was likewise demonstrated to be galactosyl transferase I [6]. This illustrates the huge potential of using *C. elegans* as a model experimental organism for mammalian glycobiological research.

Hirabayashi et al. [7] by use of glyco-catch and frontal affinity lectin chromatography are attempting to define the glycome of *C. elegans* in terms of the gene encoding the glycoprotein, the molecular mass of derived glycopeptides, the retention position of the glycopeptides in two-dimensional HPLC experiments and their dissociation constants (K_d) from a set of lectins. Preliminary data have demonstrated that Asn^{575} of a 1589-amino-acid type-II membrane protein cloned in the cosmid F54F11.2 is glycosylated. While it is acknowledged that the application of the techniques detailed above will produce valuable data, two fundamental problems could inhibit the full deci-

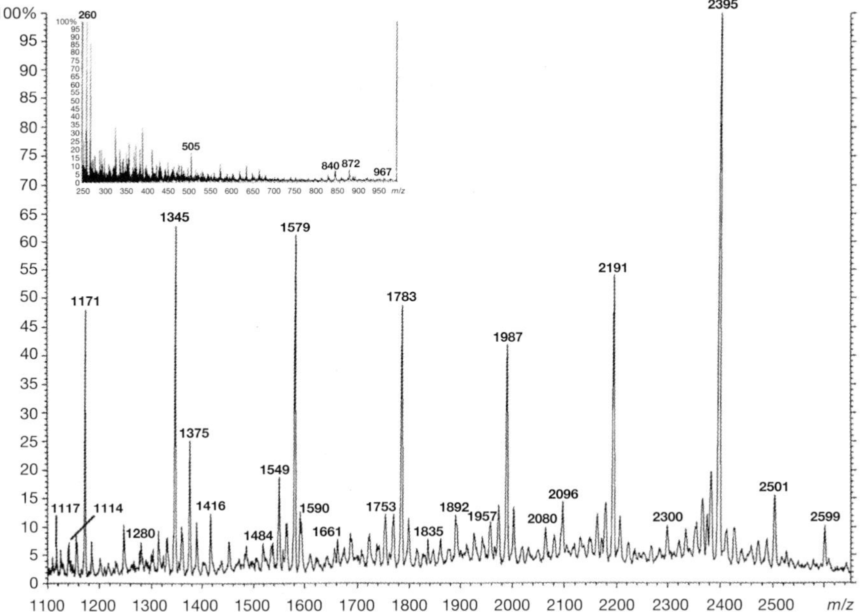

Figure 1 FAB–MS spectrum of permethylated peptide N-glycosidase-F (PNGase-F)-released N-glycans from *C. elegans.* The N-glycans of *C. elegans* were released from tryptic glycopeptides by digestion with PNGase-F, separated from protein by Sep-Pak purification and permethylated. The derivatized glycans were purified on Sep-Paks and the 50% (v/v) aqueous acetonitrile fraction was screened by FAB–MS. The low-mass region of the spectrum is included as an insert.

phering of the *C. elegans* genome. The first is that by the use of lectins to initially select the glycoproteins for analysis, large numbers of glycoproteins that carry unusual structures not bound by lectins that are available at present will be missed. Secondly, no molecular structural data are obtained. Therefore, the applications of structural techniques such as MS and NMR will be vital to fully define the *C. elegans* glycome.

Over the last few years our group at Imperial College London has been involved in the structural characterization of protein glycosylation in human and animal parasitic nematodes, as glycoconjugates are increasingly implicated in immune responses to parasitic infection [8–10]. We are using a similar, high-sensitivity MS strategy to investigate protein glycosylation in *C. elegans* (Figure 1).

Fast atom bombardment (FAB)–MS of *C. elegans* N-glycans released from detergent extracts by peptide N-glycosidase (PNGase) F and PNGase-A

Mixed-stage, wild-type *C. elegans* (strain N2, Bristol) were used in all experiments. As we have demonstrated previously as part of our characterization of N-glycosylation in *Haemonchus contortus*, the use of sequential

Table 1 Assignments of molecular and fragment ions observed in FAB spectra of permethylated N-glycans of *C. elegans*. Fuc, fucose; Hex, hexose; HexNAc, *N*-acetylhexasomine.

Signal (*m*/*z*)	Assignment
260	$HexNAc^+$
505	$HexNAc_2{}^+$
872	$Hex_3HexNAc^+$
967	$Hex_2HexNAc_2+Na^+$
1117	$Hex_3HexNAc_2{}^+$
1141	$FucHex_2HexNAc_2+Na^+$
1171	$Hex_3HexNAc_2+Na^+$
1280	$Hex_5HexNAc^+$
1315	$Fuc_2Hex_2HexNAc_2+Na^+$
1345	$FucHex_3HexNAc_2+Na^+$
1375	$Hex_4HexNAc_2+Na^+$
1416	$Hex_3HexNAc_3+Na^+$
1484	$Hex_6HexNAc^+$
1519	$Fuc_2Hex_3HexNAc_2+Na^+$
1549	$FucHex_4HexNAc_2+Na^+$
1579	$Hex_5HexNAc_2+Na^+$
1590	$FucHex_3HexNAc_3+Na^+$
1661	$Hex_3HexNAc_4+Na^+$
1693	$Fuc_3Hex_3HexNAc_2+Na^+$
1723	$Fuc_2Hex_4HexNAc_2+Na^+$
1753	$FucHex_5HexNAc_2+Na^+$
1783	$Hex_6HexNAc_2+Na^+$
1892	$Hex_8HexNAc^+$
1835	$FucHex_3HexNAc_4+Na^+$
1897	$Fuc_3Hex_4HexNAc_2+Na^+$
1927	$Fuc_2Hex_5HexNAc_2+Na^+$
1957	$FucHex_6HexNAc_2+Na^+$
1987	$Hex_7HexNAc_2+Na^+$
2071	$Fuc_4Hex_4HexNAc_2+Na^+$
2080	$FucHex_3HexNAc_5+Na^+$
2096	$Hex_9HexNAc^+$
2101	$Fuc_3Hex_5HexNAc_2+Na^+$
2191	$Hex_8HexNAc_2+Na^+$
2275	$Fuc_4Hex_5HexNAc_2+Na^+$
2300	$Hex_{10}HexNAc^+$
2395	$Hex_9HexNAc_2+Na^+$
2599	$Hex_{10}HexNAc_2+Na^+$

digestion of extracted glycopeptides with PNGase-F and PNGase-A allows a partial fractionation of released N-glycans on the basis of whether they have a fucose attached to the 3-position of the asparagine-linked *N*-acetylglucosamine (GlcNAc) residue [11,12]. The released glycans were permethylated and analysed by FAB–MS. The spectrum of the PNGase-F-released glycans (Figure 1 and Table 1) is dominated by signals consistent with high-mannose

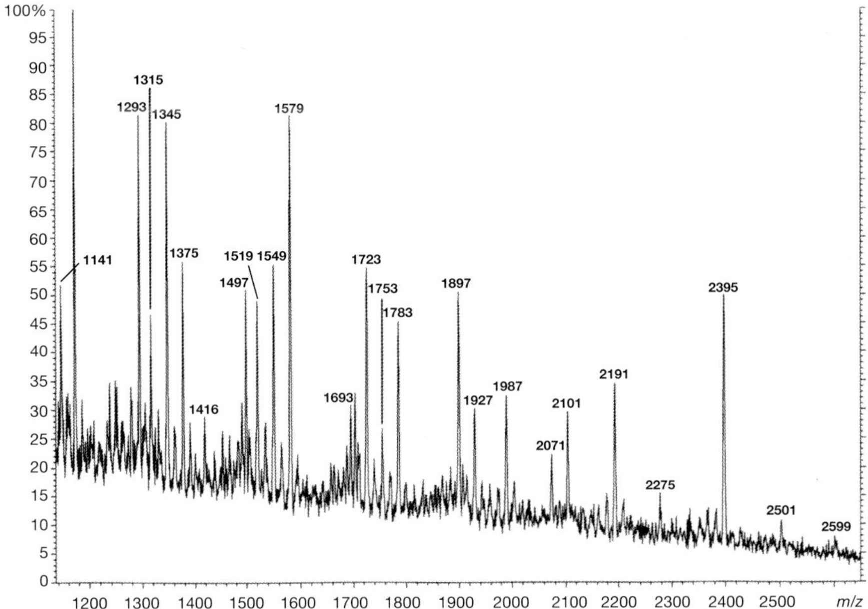

Figure 2 FAB–MS spectrum of permethylated PNGase-A-released N-glycans from *C. elegans*. The N-glycans of *C. elegans* were released from tryptic glycopeptides by digestion with PNGase-A after initial PNGase-F digestion, separated from protein by Sep-Pak purification and permethylated. The derivatized glycans were purified on Sep-Paks and the 50% (v/v) aqueous acetonitrile fraction was screened by FAB–MS.

structures (*m/z* 1579–2395, $Hex_{5-9}HexNAc_2$). Interestingly, a signal at *m/z* 2599 is also present, which is consistent with a composition of $Hex_{10}HexNAc_2$. A-type fragment ions that are compatible with these structures are also observed (*m/z* 1280–2300, $Hex_{5-10}HexNAc^+$). Signals that are consistent with small, truncated N-glycans are also abundant (*m/z* 967–1375, $Fuc_{0-1}Hex_{2-4}HexNAc_2$). One of the most striking features of the spectrum is the relatively low abundance of complex type N-glycans. Signals are present whose compositions are consistent with structures bearing short complex-type antennae (*m/z* 1416–2080, $Hex_3HexNAc_3$–$Fuc_1Hex_3HexNAc_5$), whose presence is confirmed by A-type fragment ions at *m/z* 260 $HexNAc^+$ and *m/z* 505 $HexNAc_2{}^+$ [Figure 1 (insert) and Table 1].

The spectrum of the PNGase-A-released glycans (Figure 2 and Table 1) also contains signals corresponding to high mannose and small amounts of complex N-glycans. Their presence is indicative of incomplete PNGase-F digestion. However, interestingly a new family of signals (*m/z* 1315–2275, $Fuc_{2-4}Hex_{2-5}HexNAc_2$) is also present which were not observed in the PNGase-F spectrum.

Linkage analysis

C. elegans glycans that were released by PNGase-F and PNGase-A were subjected to linkage analysis (Tables 2 and 3). These results are fully consistent with high-mannose and truncated structures being the major constituents of the N-glycan population, as terminal mannose is the most abundant linked monosaccharide. The low abundance of complex structures is indicated by the presence of very small amounts of terminal GlcNAc, 2,4-mannose and 2,6-mannose. However, the presence of 3,4,6-mannose does indicate that some of the terminal GlcNAc is bisecting. Fucose, especially in the PNGase-A-released glycans, is a major terminal sugar. The presence of 4,6-GlcNAc in the PNGase-F-released glycans and 3,4-GlcNAc and 3,4,6-GlcNAc in the PNGase-A-released glycans is consistent with core-linked fucose. Interestingly, a small proportion of 2-fucose is present in addition to the terminal fucose residue. Galactose is also a non-reducing terminus and, together with 2-galactose, it is more abundant in the PNGase-A-released glycans than in PNGase-F-released glycans.

Exo-glycosidase digestions

In order to define anomeric configuration as well as provide additional structural information, the PNGase-F- and PNGase-A-released glycans were subjected to digestion with jack bean α-mannosidase. Treatment with this enzyme causes a decrease in the intensity of signals at m/z 1579, 1783, 1987, 2191, 2395 and 2599 ($Hex_{5-10}HexNAc_2$); strong signals at m/z 763, 937 and 967 ($Fuc_{0-1}Hex_{1-2}HexNAc_2$) are predominant. In addition, linkage analysis indicates a decrease in the incidence of terminal mannose, 2-linked mannose and

Table 2 GC–MS analysis of partially methylated alditol acetates obtained from the PNGase-F-released N-glycans of *C. elegans*

Elution time (min)	Characteristic fragment ions	Assignment	Relative abundance
17.10	115, 118, 131, 162, 175	Terminal fucose	0.07
18.48	130, 131, 162, 175	2-Linked fucose	0.01
18.73	102, 118, 129, 145, 161, 205	Terminal mannose	1.00
19.00	102, 118, 129, 145, 161, 205	Terminal galactose	0.05
19.95	129, 130, 161, 190	2-Linked mannose	0.33
20.25	129, 130, 161, 190	2-Linked galactose	0.03
20.41	102, 118, 129, 162, 189, 233	6-Linked mannose	0.02
21.24	130, 190, 233	2,4-Linked mannose	0.01
21.67	129, 130, 189, 190	2,6-Linked mannose	0.01
21.82	118, 129, 189, 234	3,6-Linked mannose	0.38
22.30	118, 129, 139	3,4,6-Linked mannose	0.05
22.87	117, 159, 203, 205	Terminal GlcNAc	0.03
23.80	117, 159, 233	4-Linked GlcNAc	0.44
25.19	117, 159, 261	4,6-Linked GlcNAc	0.02

Table 3 GC–MS analysis of partially methylated alditol acetates obtained from the PNGase-A-released N-glycans of *C. elegans*

Elution time (min)	Characteristic fragment ions	Assignment	Relative abundance
17.10	115, 118, 131, 162, 175	Terminal fucose	0.53
18.45	130, 131, 162, 175	2-Linked fucose	0.07
18.68	102, 118, 129, 145, 161, 205	Terminal mannose	1.00
18.99	102, 118, 129, 145, 161, 205	Terminal galactose	0.31
19.92	129, 130, 161, 190	2-Linked mannose	0.28
20.24	129, 130, 161, 190	2-Linked galactose	0.20
21.79	118, 129, 189, 234	3,6-Linked mannose	0.15
22.27	118, 129, 139	3,4,6-Linked mannose	0.21
23.79	117, 159, 233	4-Linked GlcNAc	0.25
24.69	117, 159, 346	3,4-Linked GlcNAc	0.11
26.05	117, 159	3,4,6-Linked GlcNAc	0.05

3,6-linked mannose and an increase in the incidence of 6-linked mannose. Taken together, these results confirm the assignment of high-mannose-type structures to the signals at *m/z* 1579, 1783, 1987, 2191, 2395 and 2599 ($Hex_{5–10}HexNAc_2$). The jack bean α-mannosidase also decreases the intensity of the truncated signals at *m/z* 967–1375 ($Fuc_{0–1}Hex_{2–4}HexNAc_2$) and the complex signals at *m/z* 1416 and 1590 ($Fuc_{0–1}Hex_3HexNAc_3$), which produce new signals at *m/z* 1212 and 1386 ($Fuc_{0–1}Hex_2HexNAc_3$). The removal of high-mannose signals, which dominated both the PNGase-F and PNGase-A spectra, by jack bean α-mannosidase digestion allows clearer identification of more minor species. In the spectrum of PNGase-F-digested glycans, complex structures (*m/z* 1416–2080, $Hex_3HexNAc_3$–$Fuc_1Hex_3HexNAc_5$) and residual truncated structures (*m/z* 1141–1927, $Fuc_1Hex_2HexNAc_2$–$Fuc_2Hex_5HexNAc_2$) are now more apparent. The α-mannosidase-digested, PNGase-A-released glycan spectrum is now dominated by a series of highly fucosylated, truncated glycans (*m/z* 1111–2275, $Fuc_2HexHexNAc_2$–$Fuc_4Hex_5HexNAc_2$).

Quadrupole–time-of-flight (Q–TOF) collision-activated dissociation (CAD)–tandem MS (MS/MS) of *C. elegans* N-glycans

The ability of the Q–TOF mass spectrometer to perform CAD–MS/MS fragmentation of the molecular ions of N-glycans is proving invaluable in the assignment of detailed structures. Figure 3 shows the Q–TOF CAD–MS/MS spectrum for *m/z* 873, the $[M+2Na]^{2+}$ doubly charged molecular ion consistent with the composition $Fuc_2Hex_4HexNAc_2$, which was released by PNGase-A after an initial digestion with PNGase-F and has been digested with α-mannosidase. The main conclusion that can be drawn from these data is that there are a least two different isoforms that can be identified from structurally informative fragment ions. As depicted in Figure 3(A), the key fragment ions at *m/z* 648, 1116 and 1099 indicate that both fucose residues can

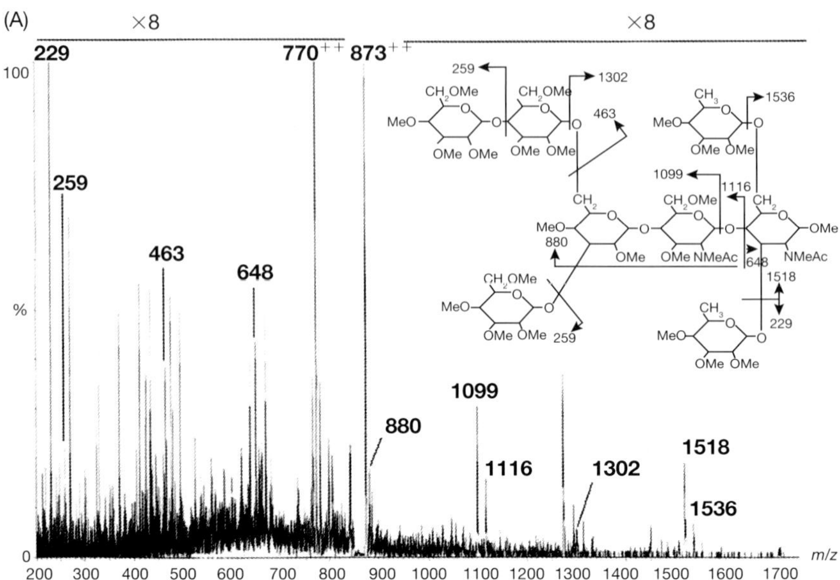

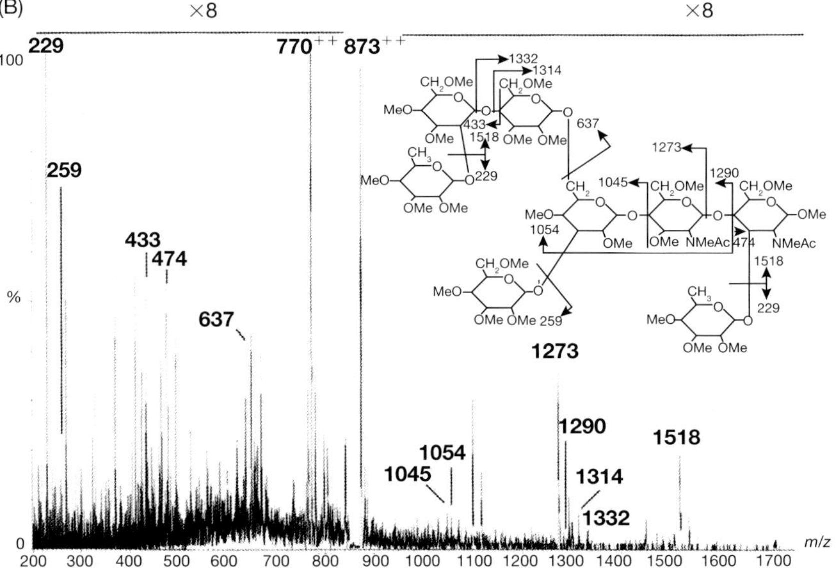

Figure 3 Q–TOF CAD-MS/MS mass spectrum of permethylated $Fuc_2Hex_4HexNAc_2$. The N-glycans of *C. elegans* were released from tryptic glycopeptides by digestion with PNGase-A after an initial PNGase-F digest, separated from protein by Sep-Pak purification and digested with α-mannosidase prior to permethylation. (**A**) and (**B**) show the fragmentation of the two different isoforms of $Fuc_2Hex_4HexNAc_2$ in Q–TOF CAD–MS/MS. The ion at *m*/*z* 770 is a doubly charged fragment ion of *m*/*z* 1518.

be linked to the reducing end GlcNAc of the chitobiose core. In addition, the fragment ion at *m/z* 880 verifies that two hexose residues are on the 6-arm of the β-linked mannose. Figure 3(B) clearly illustrates that within the same spectrum, an additional set of fragment ions is also present, which allows the identification of a second isoform. The key fragment ions at *m/z* 474, 1290 and 1273 indicate that only a single fucose residue is linked to the reducing end GlcNAc residue of the chitobiose core. The ions at *m/z* 433, 1332 and 1314 indicate that a fucose can be linked to a terminal hexose, while the ions at *m/z* 637 and 1054 confirm that the fucosylated hexose is the terminal non-reducing residue of two hexose residues on the 6-arm of the β-linked mannose. A third, minor isoform which has the fucose residue linked to the 6-position of the reducing end GlcNAc instead of the 3-position indicated in Figure 3(A), also probably exists. This isoform should have been removed by the initial PNGase-F digest but a small fraction will remain, which can subsequently be removed by PNGase-A. Experiments are under way to extend our Q-TOF analysis of the more highly fucosylated *C. elegans N*-glycans.

Identification of natural O-methyl groups on N-glycans from *C. elegans*

In our previous characterization of the O-glycans of the parasitic nematodes *Toxocara canis* and *T. cati* [13], the major structures contained natural O-methyl groups. Therefore, an experimental strategy based on the use of perdeuteromethyl derivatization of PNGase-A- and PNGase-F-released glycans was instigated. The presence of a natural O-methyl group is indicated by a molecular ion signal that is 3 mass units below that expected from a fully deuteromethylated glycan. Figure 4 shows a partial FAB–MS spectrum of perdeuteromethylated PNGase-A-released glycans. Four molecular ions can be seen at *m/z* 1795, 1861, 1977 and 2008, which is consistent with the compositions $Fuc_2Hex_4HexNAc_2$, $Hex_6HexNAc_2$, $Fuc_3Hex_4HexNAc_2$ and $Fuc_2Hex_5HexNAc_2$. Signals that are 3 mass units below the molecular ions of $Fuc_2Hex_4HexNAc_2$ and $Fuc_2Hex_5HexNAc_2$ indicate the presence of one natural O-methyl group, while signals that are 3 and 6 mass-units below the molecular ion of $Fuc_3Hex_4HexNAc_2$ indicate the presence of up to two natural O-methyl groups. Interestingly no mass reduction is observed for the molecular ion of $Hex_6HexNAc_2$, which indicates the absence of natural O-methyl groups in this component. These results indicate that the residue most likely to be substituted with the natural O-methyl group is fucose. Preliminary Q–TOF CAD–MS/MS experiments on perdeuteromethylated glycans appear to confirm this conclusion.

Structural characterization of phosphorylcholine (PC)-substituted glycans

PC has been shown to be a common substitution in several nematode glycoconjugates [14]. In our previous work on filarial nematodes, we

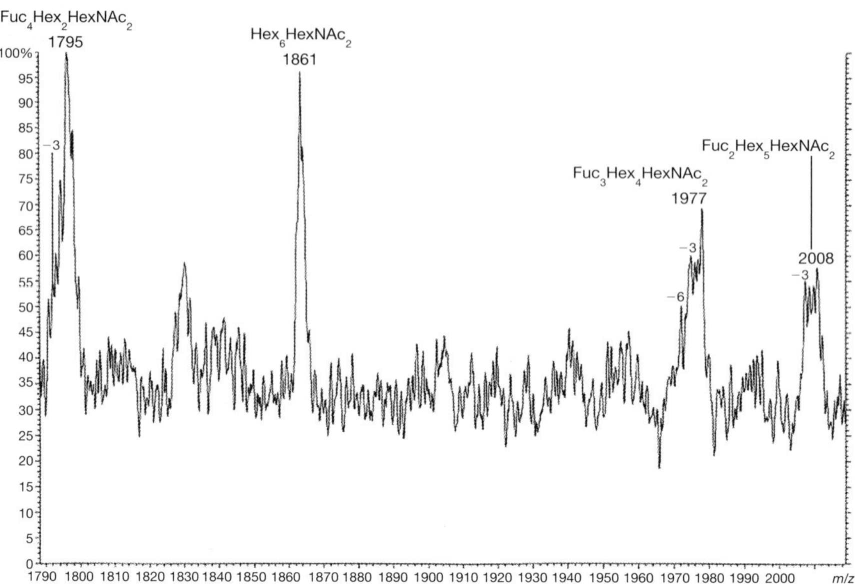

Figure 4 FAB–MS mass spectrum of perdeutromethylated PNGase-A-released N-glycans from *C. elegans*. The *N*-glycans of *C. elegans* were released from tryptic glycopeptides by digestion with PNGase-A after an initial PNGase-F digestion, separated from protein by Sep-Pak purification and perdeutromethylated. The derivatized glycans were purified on Sep-Paks and the 50% (v/v) aqueous acetonitrile fraction was screened by FAB–MS.

have demonstrated that FAB–MS analysis of perdeuteroacetylated derivatives provides a sensitive means of detecting low molecular-mass glycans that are substituted with PC [15,16]. PNGase-F-released glycans were perdeuteroacetylated and analysed by FAB–MS. Data are consistent with the compositions $Fuc_{0-1}Hex_3HexNAc_2$, $Fuc_{0-1}Hex_3HexNAc_{3-5}$ and $PCFuc_{0-1}Hex_3HexNAc_{3-5}$.

Because of the zwitterionic nature of PC, it has proven difficult to define its linkage position to nematode N-glycans. Therefore we have devised a new experimental technique based on the ability to perform MS/MS and even MS/MS/MS on the Q–TOF mass spectrometer. PC-containing deuteroacetylated glycans were analysed by Q–TOF–MS. The $[M+2H]^{2+}$ doubly charged molecular ion that is consistent with the composition $PCFuc_1Hex_3HexNAc_3$ at *m/z* 1118 was selected for CAD–MS/MS. Structurally informative fragment ions are observed at *m/z* 282 Fuc^+ (signals at *m/z* 219 and 156 are formed by sequential elimination of deuteroacetic acid), *m/z* 343 Hex^+ and at *m/z* 459 $PCHexNAc^+$. The presence of the ion at *m/z* 459 $PCHexNAc^+$ and lack of an A-type ion at *m/z* 339 $HexNAc^+$ unambiguously confirm that the non-reducing HexNAc residue is the site of PC substitution. By increasing the cone voltage to enhance cone fragmentation in the initial MS experiment the A-type fragment ion at *m/z* 459 ($PCHexNAc^+$) can be selected for CAD–MS/MS (i.e. an MS/MS/MS experiment). This approach in effect enables us to perform an MS/MS/MS experiment. The Q–TOF CAD–MS/MS

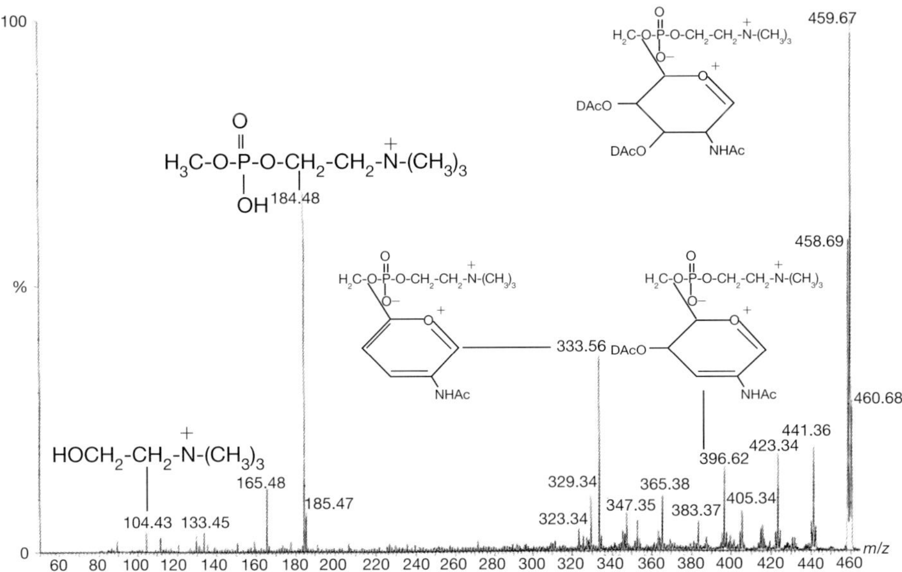

Figure 5 Q-TOF CAD-MS/MS mass spectrum of perdeutroacetylated PCHexNAc$^+$. The N-glycans of *C. elegans* were released from tryptic glycopeptides by digestion with PNGase-F, separated from protein by Sep-Pak purification and perdeutroacetylated. The A-type fragment ion PCHexNAc$^+$ was induced by cone voltage fragmentation and selected for Q–TOF CAD–MS/MS. Structurally informative fragment ions are indicated on the figure.

spectrum of *m/z* 459 PCHexNAc$^+$ is shown in Figure 5. Key fragment ions are observed at *m/z* 104 (choline), 184 (PC), 396 and 333. The fragment ions at *m/z* 396 and 333 are formed by sequential elimination of deuteroacetic acid from the C3 and C4 positions of the HexNAc residue. In addition, the lack of a fragment ion at *m/z* 276, which would have been formed by the elimination of a PC group from C3 or C4, provides strong evidence that the PC group is substituted on the 6-position of the non-reducing HexNAc.

Treatment of PNGase-F-released glycans from *C. elegans* with hyrdrofluoric acid, which we have previously demonstrated removes the PC functional group [15,16], caused a relative increase in the abundance of structures of composition $Hex_3HexNAc_3$-$Hex_3HexNAc_7$, together with an increase in the abundance of the fragment ions at *m/z* 260 HexNAc$^+$, *m/z* 505 $HexNAc_2^+$ and the production of small amounts of a new fragment ion at *m/z* 750 $HexNAc_3^+$. Linkage analysis also revealed an increase in the relative abundance of terminal GlcNAc. Digestion of the hydrofluoric acid-treated glycans with β-hexosaminidase caused a decrease in the abundance of structures of composition $Hex_3HexNAc_{3-7}$ together with an increase in the abundance of an ion at *m/z* 1171 ($Hex_3HexNAc_2$). Linkage analysis also revealed a decrease in the relative abundance of terminal GlcNAc. Taken together, these data reveal that in *C. elegans* PC is attached to the 6-position of terminal β-linked GlcNAc residues.

Structural conclusions

Four main points can be deduced from these results (summarized in Figure 6). First, high-mannose structures are by far the most abundant class of N-glycans expressed by *C. elegans*. Secondly, *C. elegans* expresses novel, highly fucosylated, truncated structures. Thirdly, only low levels of complex type N-glycans are expressed. Fourthly, *C. elegans* expresses complex N-glycans that are substituted with PC linked to the 6-position of terminal GlcNAc residues at the non-reducing ends of antennae.

The dominance of high-mannose structures has also been reported in the previous structural characterization of *C. elegans* N-glycans. Guérardel et al. [17], using NMR and MS, defined $Man_{7-9}GlcNAc_2$ structures. Altmann et al. [18], also using MS techniques, demonstrated the presence of large amounts of $Hex_{5-10}HexNAc_2$ structures. The structures were sensitive to digestion with α-mannosidase and endoglycosidase H, which confirmed that they are high-

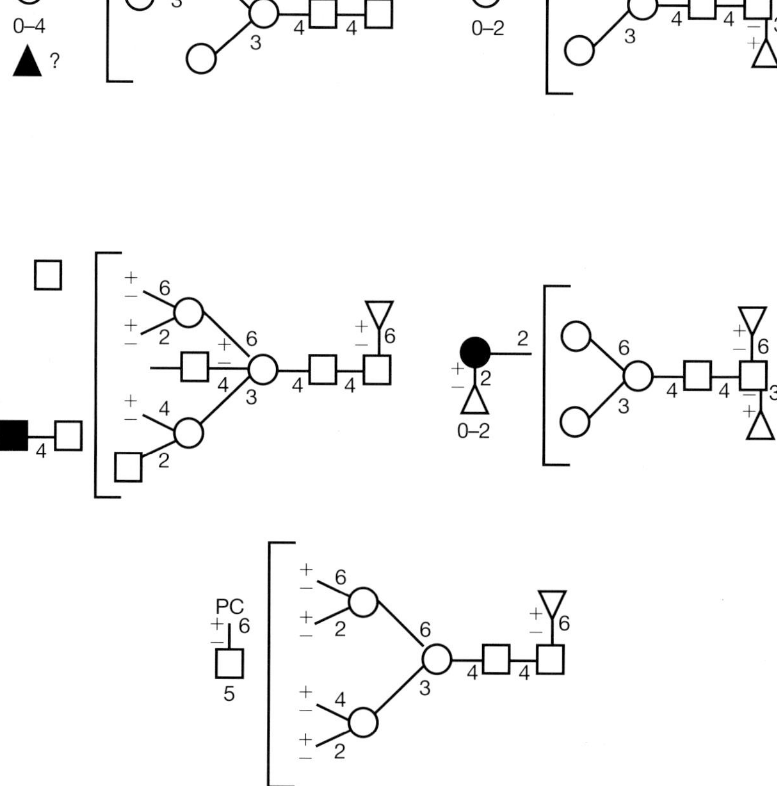

Figure 6 Structural summary of the N-glycans of *C. elegans*. ●, Galactose; ○, mannose; ▲, glucose; □, GlcNAc; ■, HexNAc; △, fucose/MeFuc; PC, phosphorylcholine.

mannose glycans. Interestingly, after α-mannosidase digestion, a small amount of $Hex_6HexNAc_2$ remained, which led the authors to conclude that the $Hex_{10}HexNAc_2$ had a composition of $Glc_1Man_9GlcNAc_2$.

Potentially the most interesting structures, in terms of their novelty, are the highly fucosylated, truncated structures ($Fuc_{2–4}Hex_{2–5}HexNAc_2$) which also are substituted by natural O-methyl groups. Again, similar compositions were observed by Altmann et al. [18]. They concluded that all the hexoses in these structures were mannose residues, that O-methyl groups were substituted on the 2-position of fucose and the 3- and 4-position of mannose and that resistance of the more highly fucosylated species to α-mannosidase digestion was due to direct linkage of fucose on to mannose. Our results lead us to different conclusions; the linkage data indicate that galactose at the non-reducing termini and 2-galactose are more abundant in PNGase-A-released glycans than in PNGase-F-released glycans. Acid hydrolysis of fucose residues from methylated glycans followed by redeutromethylation and linkage analysis indicated the incorporation of a deuteromethyl group at the 2-position of galactose, but not mannose (results not shown). Therefore fucose linked to the 2-position of galactose is a major terminal structure in the PNGase-A-released glycans. In addition, the presence of a large amount of 2-linked mannose after α-mannosidase digestion would indicate that the galactose is linked to the 2-position of mannose (data not shown). The presence of galactose and 2-fucosylated galactose explains the resistance of the highly fucosylated truncated structures to α-mannosidase. Q–TOF CAD–MS/MS fragment ion data of the $Fuc_2Hex_4HexNAc_2$ species confirms the presence of a fucose linked to a non-reducing terminal hexose in one glycoform, but also the ability to difucosylate the reducing end GlcNAc residue of the core (which is confirmed by the presence of 3,4-GlcNAc and 3,4,6-GlcNAc in the linkage analysis data). Truncated glycans with monofucosylated and difucosylated cores are expressed in *C. elegans*, as indicated by the presence of $Fuc_{1–2}Man_1GlcNAc_2$ species after α-mannosidase digestion. The Q–TOF CAD–MS/MS fragment ion data of the $Fuc_2Hex_4HexNAc_2$ species, which identified three potential iosforms of this composition, indicate the great heterogeneity of these structures due to the multiple potential sites of fucosylation. This heterogeneity is multiplied when the potential to incorporate O-methylated fucose is also taken into account.

In our previous characterizations of PC-substituted N-glycans [15,16], we were unable to definitively state the position of PC substitution. By use of a Q–TOF CAD–MS/MS strategy we have now been able to establish that PC is substituted on to the 6-position of the non-reducing GlcNAc of complex N-glycans.

O-linked glycosylation and other *C. elegans* glycoconjugates

C. elegans also appears to express unusual O-linked glycans. Using NMR and MS, Guérardel et al. [17] sequenced seven O-glycans, five of which had a type-1 core. The cores were extended by the addition of β-linked glucosamine, galactose, glucuronic acid and GlcNAc. In addition, a larger, nine-residue structure had a reducing end GlcNAc residue as well as α-linked fucose and 2-O-methylated-fucose. A pentameric glycosaminoglycan-like structure that contains the conserved GlcAβ1–3Galβ1–3Galβ1–4Xyl core sequence with an additional non-reducing end β1–4-linked *N*-acetylgalactosamine (GalNAc) was also characterized. Yamada et al. [19] demonstrated that adult *C. elegans* contain considerable amounts of non-sulphated chondroitin, less heparan sulphate and no detectable hyaluronate. Analysis of the sulphation pattern of the heparan sulphate revealed the presence of 2-N-sulphated glucosamine, 6-O-sulphated glucosamine and 2-O-sulphated uronate.

C. elegans also expresses both neutral and charged PC-substituted glycosphingolipids. The three neutral glycolipids are relatively simple, comprising Glcβ1Cer, Manβ1–4Glcβ1Cer and GlcNAcβ1–3Manβ1–4Glcβ1Cer [20], whereas the three major PC-substitued glycosphingolipids are more complex, comprising GalNAcβ1–4[PC]GlcNAcβ1–3Manβ1–4Glcβ1Cer, Galα1–3GalNAcβ1–4[PC]GlcNAcβ1–3Manβ1–4Glcβ1Cer and Galβ1–3 Galα1–3GalNAcβ1–4[PC]GlcNAcβ1–3Manβ1–4Glcβ1Cer [21]. It is interesting to note that in all cases the PC was linked to GlcNAc as we have described in the N-linked glycans of *C. elegans*.

How do the glycan structures of *C. elegans* reflect the sequenced genome?

The sequencing of the *C. elegans* genome was completed in 1998 [1] and represents a major information resource to be exploited by glycobiologists. However, it is only now that detailed glycan structural data are becoming available that the full potential of this resource can be exploited. This situation is probably most clearly illustrated by *C. elegans* fucosyltransferases. By using a genomics approach, Oriol et al. [22] predicted the presence of 18 different putative fucosyltransferases, including members of the α-2, α-3 and α-6 families. As has been described above our structural characterization of the N-linked glycans of *C. elegans* revealed a family of novel highly fucosylated structures that contained α-2-, α-3- and α-6-linked fucose residues. Therefore the initial predictions made by analysing the *C. elegans* genome have been vindicated by structural analysis. To date, only a single α1-3fucosyltransferase has been cloned [23] therefore *C. elegans* remains a tremendous potential source of fucosyltransferases to be exploited by glycobiologists. In addition, three N-acetylglucosaminyltransferase I [24] and five N-acetylgalactosamine:polypeptide N-acetylgalactosaminyltransferases [25] have also been cloned from *C. elegans*, which are fully consistent with the structure of the glycans that are detailed above.

It is interesting to note that structural characterization of glycolipids from the parasitic nematodes *Ascaris suum* [26] and *Onchocerca volvulus* [27] also revealed that PC is attached to the 6-position of GlcNAc as we have described in *C. elegans*. Therefore there could be a conservation of the PC transferase enzymes across nematode species, which could be cloned and characterized from *C. elegans*. As mammalian host glycoconjugates do not contain PC, the nematode PC transferase could be a potential chemotherapeutic target.

C. elegans also appears to be a rich source of carbohydrate-binding proteins. Two galectins have been purified from *C. elegans*. The first, a 32-kDa protein, was the first example of a galectin in an invertebrate and also the first tandem-repeat-type structure in which both N-terminal and C-terminal lectin domains have been shown to bind galactose [28,29]. The second, a 16-kDa protein, was initially assumed to be a degradation product of the larger 32-kDa species but has since been demonstrated to be a novel prototype galectin [30]. The structural data presented should also provide an insight into the carbohydrate ligands of the *C. elegans* galectins. Arata et al. [31] have used a frontal affinity chromatography system to investigate the sugar-binding properties of the two lectin domains of the 32-kDa tandem-repeat-type galectin. They demonstrated that both lectin domains have affinity for complex N-glycans with *N*-acetyl-lactosamine-containing antennae. However, we could find no evidence for *N*-acetyl-lactosamine-containing complex N-glycans in *C. elegans*. Interestingly, Arata et al. [31] also demonstrated that both the individual lectin domains and the intact tandem galectin showed strong affinity for the H-antigen structure (Fucα1–2Galβ1–3GlcNAc). Therefore, the *in vivo* ligands for this *C. elegans* galectin could be the novel fucose-rich, truncated structures that contain Fucα1–2Gal1–2Man sequences, which we have described.

In addition, Drickamer and Dodd [32], using an *in silico* approach, identified 183 C-type lectin-like domains in at least 125 *C. elegans* proteins. However, in contrast to the large number of membrane-bound mammalian C-type lectins, all the *C. elegans* proteins with C-type lectin-like domain that resemble carbohydrate-recognition domains are found in proteins lacking membrane anchors and are therefore likely to be secreted. Our structural data indicate a very low level expression of complex type N-glycan structures in *C. elegans*. Taken together, these results indicate that *C. elegans* may lack many endocytic and cell adhesion functions associated with interactions between complex N-linked glycans and C-type lectins in higher organisms. Drickamer and Dodd hypothesized that these secreted proteins could perform functions that are similar to those of soluble mannose-binding proteins in the innate immune response of *C. elegans*.

As with all other structural studies of nematode glycosylation that have been performed to date, no sialated structures were detected. This is fully consistent with the findings that *C. elegans* lacks the enzyme genes required for sialic acid biosynthesis and also lacks any siglec homologues [33]. This lack of sialic acid on nematode glycans could explain the large numbers of fucosylated structures and fucosyltransferases, as in mammalian glycans fucose and sialic acid compete to cap non-reducing antennae [22].

Glycomes of other model organisms

As well as *C. elegans*, glycobiologists have exploited other model organisms, most notably *D. discoideum, S. cerevisiae* and *D. melanogaster.* A comprehensive description of the glycobiology of these organisms is beyond the scope of this chapter; however, the interested reader is directed to the several excellent reviews listed in each section.

D. discoideum

The haploid, eukaryotic amoeba *D. discoideum* usually exists as a single cellular species that phagocytoses bacteria. However, in times of food shortage a complex and synchronous morphogenesis is initiated, which leads to the production of multicellular species. This elaborate life cycle has made *D. discoideum* an attractive model organism for cell and developmental biologists [34]. The sequence of the *D. discoideum* genome is expected to be completed by mid 2003 [35].

D. discoideum expresses mostly high-mannose-type N-glycan structures, which may be substituted with GlcNAc, xylose and mannose 6-phosphate residues [34]. In addition, antibody-binding experiments indicate the presence of both Fucα1–6 and Fucα1–3 core fucosylation [36,37]. *D. discoideum* also expresses some unusual forms of O-glycosylation in proteins. These include the addition of GlcNAcα 1-phosphate to serine residues [38], Fucβ 1-phosphate to serine residues [37], GlcNAcα1- to serine/threonine residues [39] and a novel form of nuclear and cytoplasmic glycosylation, in which the pentasaccharide Galα1–6Galα1–3Fucα1–2Galβ1–3GlcNAc is attached to hydroxyproline [40].

D. melanogaster

The fruit-fly *D. melanogaster* is one of the most favoured model organisms used by molecular geneticists. The sequence of the *D. melanogaster* genome was completed in 2000 and is the most comprehensive genome to be annotated for any multicellular organism [41]. Two recent reviews have comprehensively catalogued the known glycan structures from *D. melanogaster* [18,42].

The N-glycan profile of *D. melanogaster* appears similar to that described for *C. elegans*, in that it also tends to produce mostly high-mannose type structures with some small complex structures and truncated structures, which can have both Fucα1–6 and Fucα1–3 attached to the GlcNAc residue at the reducing end of the chitobiose core [43,44]. *D. melanogaster* also expresses short core type-1 O-glycans, Galβ1–3GalNAc [45] and O-linked GlcNAc on nuclear proteins [46]. One of the most exciting recent discoveries in functional glycobiology was the demonstration that Fringe is an O-fucose β1-3-N-acetylglucosaminyltransferase that adds GlcNAc to O-linked fucose on epidermal growth factor-like repeats of Notch. This modulates Notch-dependent signalling pathways that establish dorso–ventral boundaries during *D. melanogaster* embryogenesis [47,48].

S. cerevisiae

The *S. cerevisiae* genome sequence was completed in 1997 [49], and the yeast is probably the best-characterized model organism in terms of its glycobiology. A detailed review of *S. cerevisiae* and other yeast N- and O-linked glycan structures was recently published by Gemmill and Trimble ([50]; other review articles on yeast glycobiology are included in the same volume). *S. cerevisiae* produces two classes of polymannan N-glycans. Secreted or cell-wall structures can contain up to 200 mannose units, the outermost of which can be phosphorylated. In contrast, intracellular glycoproteins contain much smaller $Man_{9-15}GlcNAc_2$ structures. *S. cerevisiae* also produces simple Man_{1-6} O-linked glycans which can also be phosphorylated.

S.M.H. would like to thank the Biotechnology and Biological Sciences Research Council and the Wellcome Trust for financial support, and Simon North and Maria Panico for assistance with Q–TOF–MS experiments. D.G. is the recipient of a Royal Society University Research Fellowship.

References

1. The *C. elegans* Sequencing Consortium (1998) Science **282**, 2012–2018
2. Sulston, J.E., Schierenberg, E., White, J.G. and Thomson, J.N. (1983) Dev. Biol. **100**, 64–119
3. Wood, W.B. (ed.) (1989) The Nematode *Caenorhabditis elegans.* Cold Spring Harbor Laboratory Press, Cold Spring Harbor, NY
4. Herman, T. and Horvitz, H.R. (1999) Proc. Natl. Acad. Sci. U.S.A. **96**, 974–979
5. Bulik, D.A., Wei, G., Toyoda, H., Kinoshita-Toyoda, A., Waldrip, W.R., Esko, J.D., Robbins, P.W. and Selleck, S.B. (2000) Proc. Natl. Acad. Sci. U.S.A. **97**, 10838–10843
6. Okajima, T., Yoshida, K., Kondo, T. and Furukawa, K. (1999) J. Biol. Chem. **274**, 22915–22918
7. Hirabayashi, J., Arata, Y. and Kasai, K. (2001) Proteomics **1**, 295–303
8. Dell, A., Haslam, S.M., Morris, H.R. and Khoo, K.-H. (1999) Biochim. Biophys. Acta **1455**, 353–362
9. Dell, A., Haslam, S.M. and Morris, H.R. (2001) in Parasitic Nematodes: Molecular Biology, Biochemistry and Immunology (Kennedy, M.W. and Harnett, W., eds.), pp. 285–307, CABI Publishing, Wallingford, U.K.
10. Haslam, S.M., Dell, A. and Morris, H.R. (2001) Trends Parasitol. **5**, 231–235
11. Haslam, S.M., Coles, G.C., Munn, E.A., Smith, T.S., Smith, H.F., Morris, H.R. and Dell, A. (1996) J. Biol. Chem. **271**, 30561–30570
12. Haslam, S.M., Coles, G.C., Reason, A.J., Morris, H.R. and Dell, A. (1998) Mol. Biochem. Parasitol. **93**, 143–147
13. Khoo, K.-H., Maizels, R.M., Page, A.P., Taylor, G.W., Rendell, N.B. and Dell, A. (1991) Glycobiology **1**, 163–171
14. Lochnit, G., Dennis, R.D. and Geyer, R. (2000) Biol. Chem. **381**, 839–847
15. Haslam, S.M., Khoo, K.-H., Houston, K.M., Harnett, W., Morris, H.R. and Dell, A. (1997) Mol. Biochem. Parasitol. **85**, 53–66
16. Haslam, S.M., Houston, K.M., Harnett, W., Reason, A.J. Morris, H.R. and Dell, A. (1999) J. Biol. Chem. **274**, 20953–20960
17. Guérardel, Y., Balanzino, L., Maes, E., Leroy, Y., Coddeville, B., Oriol, R. and Strecker, G. (2001) Biochem. J. **357**, 167–182
18. Altmann, F., Fabini, G., Ahorn, H. and Wilson, I.B.H. (2001) Biochimie **83**, 703–712
19. Yamada, S., Van Die., I., Van den Eijnden, D.H., Yokota, A., Kitagawa, H. and Sugahara, K. (1999) FEBS Lett. **459**, 327–331
20. Gerdt, S., Lochnit, G., Dennis, R.D. and Geyer, R. (1997) Glycobiology **2**, 265–275
21. Gerdt, S., Dennis, R.D., Borgonie, G., Schnabel, R. and Geyer, R. (1999) Eur. J. Biochem. **266**, 952–963
22. Oriol, R., Mollicone, R., Cailleau, L.B. and Breton, C. (1999) Glycobiology **9**, 323–334
23. DeBose-Boyd, R.A., Nyame, A.K. and Cummings, R.D. (1998) Glycobiology **8**, 905–917
24. Chen, S.H., Zhou, S.H., Sarkar, M., Spence, A.M. and Schachter, H. (1999) J. Biol. Chem. **274**, 288–297
25. Hagen, F.K. and Nehrke, K. (1998) J. Biol. Chem. **273**, 8268–8277

26. Lochnit, G., Dennis, R.D., Ulmer, A.J. and Geyer, R. (1998) J. Biol. Chem. **278**, 466–477
27. Wuhrer, M., Rickhoff, S., Dennis, R.D., Lochnit, G., Soboslay, P.T., Baumeister, S. and Geyer, R. (2000) Biochem. J. **348**, 417–423
28. Hirabayashi, J., Satoh, M., Ohyama, Y. and Kasai, K. (1992) J. Biochem. (Tokyo) **111**, 553–555
29. Hirabayashi, J., Satoh, M. and Kasai, K. (1992) J. Biol. Chem. **267**, 15485–15490
30. Hirabayashi, J., Ubukata, T. and Kasai, K. (1996) J. Biol. Chem. **271**, 2497–2505
31. Arata, Y., Hirabayashi, J. and Kasai, K.-I. (2001) J. Biol. Chem. **276**, 3068–3077
32. Drickamer, K. and Dodd, R. (1999) Glycobiology **9**, 1357–1369
33. Angata, T. and Varki, A. (2000) J. Biol. Chem. **275**, 22127–22135
34. Freeze, H.H.(1997) in Glycoproteins II (Montreuil, J., Vliegenthart, J.F.G. and Schachter, H., eds), pp. 89–122, Elsevier, Amsterdam
35. Kay, R.R. and Williams, J.G. (1999) Trends Genet. **15**, 294–297
36. Srikrishna, G., Varki, N.M., Newell, P.C., Varki, A. and Freeze, H.H. (1997) J. Biol. Chem. **272**, 25743–25752
37. Srikrishna, G., Wang, L. and Freeze, H.H. (1998) Glycobiology **8**, 799–811
38. Gustafson, G.L. and Milner, L.A. (1980) J. Biol. Chem. **255**, 7208–7210
39. Jung, E., Gooley, A.A., Packer, N.H., Slade, M.B., Williams, K.L. and Dittrich, W. (1997) Biochemistry **36**, 4034–4040
40. Teng-umnuay, P., Morris, H.R., Dell, A., Panico, M., Paxton, T. and West, C.M. (1998) J. Biol. Chem. **273**, 18242–18249
41. Rubin, G.M., Hong, L., Brokstein, P.M., Evans-Holm, M., Frise, E., Stapleton, M. and Harvey, D.A. (2000) Science **287**, 2222–2224
42. Seppo, A. and Tiemeyer, M. (2000) Glycobiology **10**, 751–760
43. Altmann, F., Staudacher, E, Wilson, I.B.H and März, L. (1999) Glycoconj. J. **2**, 109–123
44. Fabini, G., Freilinger, A., Altmann, F. and Wilson, I.B.H. (2001) J. Biol. Chem. **276**, 28058–28067
45. Kramerov, A.A., Arbatsky, N.P., Rozovsky, Y.M., Mikhaleva, E.A., Polesskaya, O.O., Gvozdev, V.A. and Shibaev, V.N. (1996) FEBS Lett. **378**, 213–218
46. Kelly, W.G. and Hart, G.W. (1989) Cell **57**, 243–251
47. Buckner, K., Perez, L., Clausen, H. and Cohen, S. (2000) Nature (London) **406**, 411–415
48. Moloney, D.J., Panin, V.M., Johnston, S.H., Chen, J., Shao, L., Wilson, R., Wang, Y., Stanley, P., Irvine, K.D., Haltiwanger, R.S. and Vogt, T.F. (2000) Nature (London) **406**, 369–375
49. Goffeau, A., Barrel, B.G., Bussey, H., Davis, R.W., Dujon, B., Feldmann, H., Galibert, F., Hoheisel, J.D., Jacq, C., Johnston, M. et al. (1997) Science, **274**, 546–567
50. Gemmill, T.R. and Trimble, R.B. (1999) Biochim. Biophys. Acta **1426**, 227–237

Biochem. Soc. Symp. **69**, 135–142
(Printed in Great Britain)

11

Custom microarray for glycobiologists: considerations for glycosyltransferase gene expression profiling

Elena M. Comelli*†‡§, Margarida Amado*†§, Steven R. Head*†§ and James C. Paulson*†§[1]

*Department of Molecular Biology, The Scripps Research Institute, 10550 North Torrey Pines Road, La Jolla, CA 92037, U.S.A., †Department of Molecular and Experimental Medicine, The Scripps Research Institute, 10550 North Torrey Pines Road, La Jolla, CA 92037, U.S.A., ‡Nestlé Research Centre, Lausanne, Switzerland, and §the DNA Array Core Facility, The Scripps Research Institute, 10550 North Torrey Pines Road, La Jolla, CA 92037, U.S.A.

Abstract

The development of microarray technology offers the unprecedented possibility of studying the expression of thousands of genes in one experiment. Its exploitation in the glycobiology field will eventually allow the parallel investigation of the expression of many glycosyltransferases, which will ultimately lead to an understanding of the regulation of glycoconjugate synthesis. While numerous gene arrays are available on the market, e.g. the Affymetrix GeneChip® arrays, glycosyltransferases are not adequately represented, which makes comprehensive surveys of their gene expression difficult. This chapter describes the main issues related to the establishment of a custom glycogenes array.

Introduction

Glycans occur in prokaryotes and eukaryotes either as co- or post-translational modifications of proteins or linked to lipids. Since they are present mainly at the cell surface, changes in their expression influence many biological events, such as embryogenesis, cell–cell interactions, cell activation and differentiation, and host–pathogen interactions [1]. The structure of a specific

[1]To whom correspondence should be addressed (e-mail jpaulson@scripps.edu).

glycoconjugate is determined by the activity of specific glycosyltransferases that, in a sequential action, add sugar moieties to the growing oligosaccharide chain. Therefore, the expression of a specific glycoconjugate does not depend on the expression of a single gene, but rather on the concerted expression of all the glycosyltransferase genes involved in its synthesis. Gene expression holds a primary role in the glycobiology field, since the action of glycosyltransferases is thought to be regulated mainly at the mRNA level. Classical techniques, such as Northern blotting, competitive reverse transcriptase–PCR and *in situ* hybridization have been extensively used in the past. However, because these methods allow analysis of only one gene at a time, several experiments are required to elucidate the whole mechanism involved in the synthesis of a single glycoconjugate. In addition, glycosyltransferases are typically expressed at very low levels and therefore high amounts of starting material (total RNA) are required when using classical techniques.

Recent developments in microarray technology have enabled analysis of gene expression of thousands of genes in parallel. In a typical microarray experiment, the sample mRNA is reverse transcribed to produce fluorescently labelled cDNA (the target). The target is hybridized to a set of DNA sequences (the probe), which are bound to a solid surface (the array). The array is then scanned, and the fluorescence intensities produced by each probe are analysed with image-quantification software. In the following sections, the implementation of this microarray technology in the glycobiology field is described.

Microarray technology from a glycobiological perspective

DNA microarray technology has developed along two main lines; the Affymetrix approach (Affymetrix, Santa Clara, CA, U.S.A.), based on microarrays of oligonucleotides synthesized *in situ* by photolithographic techniques [2] and the approach pioneered at Stanford University (Stanford, CA, U.S.A.), which involves the production of a microarray by spotting pre-synthesized PCR products [3] or oligonucleotides [4,5]. Affymetrix GeneChip® arrays are relatively easy to process and the related experimental procedures and instrumentation are well optimized; however, high costs may limit their use on a routine basis. An additional limitation of the Affymetrix GeneChip® system for glycobiologists is the absence of a substantial number of the cloned glycosyltransferase genes (only 60% of the known human glycosyltransferases genes are represented on the most recent version of the Affymetrix human GeneChip® array, U95Av2). Taking this into consideration, the construction of a custom array containing all the available glycosyltransferase genes was planned. Different methods are available to accomplish each of the steps of a microarray experiment, ranging from the printing technology and the choice of the DNA probes to the labelling of the sample [6].

Choice of the DNA printing technology

Several options are available regarding the printing technology, i.e. the devices used to deliver the DNA probes and the receiving surfaces. Both non-contact and contact printing have been used. The first of these, which is normally referred to as ink-jet printing, is carried out with a sharpened end capillary dispenser through which the solution is ejected in droplets on to the surface. The latter uses pins, which carry the solution at their tips and release it when they make contact with the surface. Glass microscope slides (25 mm × 75 mm), which are pre-treated to allow the attachment of DNA, have been the typical support used for custom microarrays. The slides can be coated with amine or polylysine in order to bind DNA through ionic interactions, or they can be aldehyde-derivatized in order to bind amino-modified DNA by covalent linkage.

For the glycogenes array, contact printing on polylysine-coated slides was chosen, using quill pins assembled in an arrayer that is based on the one used at Stanford University (http://cmgm.stanford.edu/pbrown/mguide/index.html). Such pins deliver submicrolitre volumes, resulting in spots with a diameter of approx. 150 μm. Several nanograms of DNA are deposited per spot, of which only a small fraction is retained on the surface after slide washing. The fractional occupancy of each spot of the array by the complementary target transcript present in the labelled cDNA mixture is proportional to the concentration of the transcript in the mixture [7], thus providing the means to measure the relative abundance of different transcripts.

Choice of DNA probe format

The type of arrayer used in our laboratory has been extensively employed by others to spot cDNA or PCR products (100–2000 bp in length). For the construction of the glycogenes array, the less commonly used synthetic oligonucleotides were chosen as probe sequences. The advantages of using oligonucleotides over PCR products include ease of selection of the probe sequence, and the low cost and simplicity of synthesis. Enough material can be generated in a single oligonucleotide synthesis to print thousands of arrays. In contrast, the production of cDNA clones, together with PCR amplification and purification steps, can be very expensive. In addition, oligonucleotides are inherently sequence-verified, eliminating the problem of sequence confirmation, which is necessary when using cDNA probes. Synthetic oligonucleotides (75 bp long) were chosen, based on previously published data using spotted micorarrays [5,7,8]. Stable attachment of oligonucleotides to the polylysine surface was verified by hybridizing with SYBR Green II stain (Molecular Probes, Eugene, OR, U.S.A.). Sensitivity and specificity were evaluated by calculating the ratio of specific to non-specific hybridization signals and by hybridizing the microarray with synthetic mRNAs generated by *in vitro* transcription from cloned sequences.

Choice of sequence for glycosyltransferases probes

Oligonucleotide probe sequences were selected to have similar GC content (in the range 40–52%) and low potential for hairpin and double-strand formation, and were located within 500 bp of the polyadenylated tail. This last point is of major importance, since the sample RNA will be labelled by reverse transcription using oligo(dT) primer. In the case of the glycosyltransferase genes, having the target in the 3′-untranslated region (UTR) represents an additional advantage, given the close similarity between the coding sequences of glycosyltransferase family members. However, the existence of multiple polyadenylation signals must be taken into account, since different transcripts can be differentially expressed in different tissues [9]; in this case, more than one oligonucleotide is required. Oligonucleotide probes for glycosyltransferases were designed using the Oligo 6.42 primer design software (Molecular Biology Insights, Cascade, CO, U.S.A.). Each sequence was analysed using the BLASTn algorithm (www.ncbi.nlm.nih.gov/BLAST/), and sequences with a similarity of more than 18 bp to other sequences from the same species were discarded.

Preparation of the cDNA target

The cDNA target can be obtained from the sample RNA by direct or indirect labelling. With classical direct labelling methods, the sample RNA is labelled by incorporating cyanine (Cy)-tagged dUTP or dTTP during a reverse transcription reaction that is primed with oligo(dT). The amount of fluorescent dye incorporated in the final probe will therefore depend on the length and sequence of the transcript. Usually Cy3 and Cy5 dyes are used for the two samples to be compared, and differential gene expression is then represented by the ratio of the intensity of one to the other. Alternatively, the dendrimer technology (Genisphere, Montvale, NJ, U.S.A.) allows use of a 5′-end-modified oligo(dT) primer which hybridizes, after the reverse transcription step, to a fluorescently labelled DNA dendrimer. This procedure should ensure uniform labelling of all the transcripts. Other methods, that have not been evaluated in our laboratory, are available, e.g. the incorporation of epitope tags into the sample cDNA and detection after hybridization by protein staining (for example biotin labelling followed by streptavidin staining). The starting RNA can either be total RNA prepared from a tissue sample or amplified RNA obtained by *in vitro* transcription with T7 RNA polymerase from double-stranded cDNA [10]. Since the products that result from these two methods yield cDNA sequences from opposite strands, both sense and antisense oligonucleotide probes should be printed on the slide, as was done for the glycogenes array described here. The presence of sense oligonucleotides when using amplified RNA (and antisense when using total RNA) provides additional negative controls and a valuable means of quantifying the background signal. Preliminary experiments revealed that glycosyltransferase expression levels were quite low relative to housekeeping genes. Therefore, RNA amplification is routinely used to improve sensitivity. As described above, after RNA amplification, the labelled cDNA binds to the antisense oligonucleotide probes arrayed on the glass slide.

Quantification of the signals and normalization issues

To detect the fluorescent hybridization signals, the glycogene microarrays are scanned with a Packard ScanArray 5000 (www.packardinstrument.com) at a resolution of 10 μm at the appropriate wavelengths for Cy3 and Cy5 detection. For the quantification of microarray signals, the files are analysed with ImaGene™ 4.1 software (BioDiscovery, Los Angeles, CA, U.S.A.). This software is one of the several image analysis programs designed for manual and semi-automatic spot finding and quantitation. The background signal is the average signal intensity from the 'sense' oligonucleotides of the housekeeping genes (when using amplified RNA), and this value is subtracted from the average signal intensity obtained for the antisense oligonucleotides for each target gene. A cut-off of two times the background is considered to accept or discard signals. For tissue expression comparisons in dual channel experiments, the signal between channels is normalized based on average expression of the housekeeping genes in the two samples. The ratio of intensity in the Cy5 channel to the one in the Cy3 channel is then used to evaluate the differential gene expression.

Towards a glycosyltransferase microarray

The first prototype glycogene array contains nine mouse glycosyltransferases, two neuraminidases, seven lymphocyte markers and twelve housekeeping genes, each represented by a minimum of two sense and one antisense oligonucleotide (Table 1). For these genes, the complete sequence, including the entire 3′-UTR, was available in GenBank. RNA obtained from various mouse tissues, in which the expression of the selected glycosyltransferase genes is well documented, was analysed in dual channel experiments to validate the approach. As an example, Figure 1 illustrates differential expression of glycosyltransferases between kidney and liver. Several genes were expressed at higher levels in kidney compared with liver, such as the core 2 β-1,6 N-acetylglucosaminyltransferase (C2-GnT) a sialyltransferase [α-N-acetylgalactosaminide α-2,6 sialyltransferase (ST6GalNAc)II] and neuraminidase (Neu)1, as has been reported previously [11–13]. In contrast, the expression of two sialyltransferases, β-galactoside α-2,6 sialyltransferase I (ST6Gal I) and β-galactoside α-2,3 sialyltransferase I (ST3Gal I) was higher in liver, also in accordance with former studies [14,15]. It should be noted that

Table 1 Mouse glycosyltransferases and neuraminidases represented on the glycogenes array prototype. β4 Gal-T1, β-N-acetylglucosamine β-1,4 galactosaminyltransferase 1; C1GalT1, α-N-acetylglucosamine β-1,3 galactosaminyltransferase 1; C2-GnT, core 2 β-1,6 N-acetylglucosaminyltransferase; FUT VII, fucosyltransferase VII; ST3Gal, β-galactoside α-2,3 sialyltransferase; ST6Gal I, β-galactoside α-2,6 sialyltransferase I; ST6GalNAc, α-N-acetylgalactosaminde α-2,6 sialyltransferase; Neu, neuraminidase.

Gene	GenBank accession number
β4Gal-T1	J03880
C1Gal T1	AF157962
C2-GnT	NM_010265
FUT VII	NM_013524
ST3Gal I	X73523
ST3Gal III	BC006710
ST6Gal I	D16106
ST6GalNAc I	NM_011371
ST6GalNAc II	X93999
Neu1	BC004666
Neu3	NM_016720

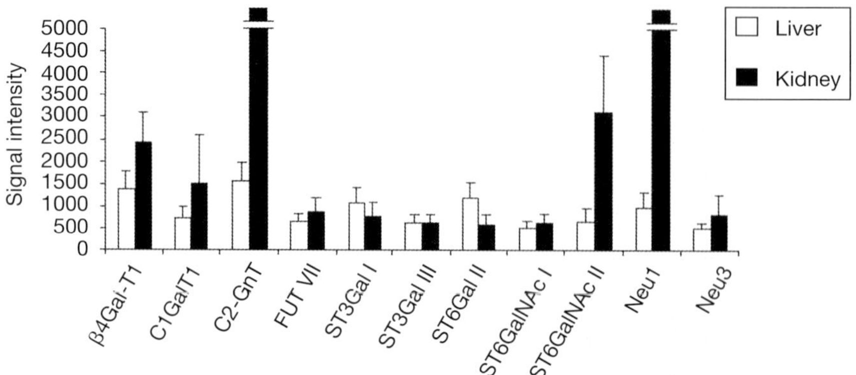

Figure 1 Differential glycosyltransferases gene expression in mouse liver and kidney. Samples of amplified RNA (3 μg) from liver and kidney were labelled with Cy3- and Cy5-dUTP respectively and were hybridized in dual channels for 12 h at 50°C in 20% formamide buffer. The plot illustrates the average signal intensities produced by each probe after background subtraction (the background was assigned as the average intensities of the sense house-keeping oligonucleotides). See Table 1 for abbreviations.

microarray data are generally effective in indicating whether a particular gene transcript level is increased or decreased across samples, but not in determining absolute expression levels. Indeed, the fluorescence produced by each hybridized cDNA target depends on its sequence and on the efficiency of the reverse transcription reaction, as mentioned above.

When planning implementation of this array to include all glycosyltransferase sequences publicly available, the main issue that needs to be addressed regards the choice of the most suitable DNA sequence to be used for probe design.

Glycosyltransferases can be classified in gene families based on either sequence identity {Carbohydrate-Active Enzymes (CAZY) server at http://afmb.cnrs-mrs.fr/~cazy/CAZY/index.html [16]} or donor substrate specificity. The number of glycosyltransferase genes cloned has increased dramatically and currently 150 human and mouse sequences are available. Examination of the databases reveals redundancy in the number of sequences for the same gene. As an example, in the CAZY database, 55 sequences (GenBank accession numbers) are listed for only 18 mouse sialyltransferases. When redundant accessions are available for a particular gene of interest, the main criterion for choosing the best sequence probe design is the selection of the one with the longest 3′-UTR. However, sometimes only part of the UTR has been deposited and in some cases only the coding sequence is present, making the design of oligonucleotides difficult (Table 2). In such situation, a probe can still be designed, but one must be bear in mind that the reverse-transcribed target may not cover it. The size of transcripts in published Northern blot experiments can provide an estimate of the risk of synthesizing a cDNA not complementary to the printed probe. The possible presence of alternative splicing in the 3′-UTR and multiple polyadenylation signals can complicate matters

Table 2 Human and mouse glycosyltransferase sequences publicly available in the CAZY and GenBank databases. Families are based on donor substrate specificity. Glycosyltransferases involved in proteoglycans synthesis are not included. Fuc, fucose; NeuAc, N-acetylneuraminic acid; Gal, galactose; GalNAc, *N*-acetylgalactosamine; Glc, glucose; GlcNAc, *N*-acetylglucosamine; Man, mannose.

Family	Species	Sequences available for probe design	Sequences with complete 3′-UTR	Sequences with incomplete or missing 3′UTR
Fuc transferases	Human	10	9	1
	Mouse	7	3	4
NeuAc transferases	Human	17	10	7
	Mouse	18	14	4
Gal transferases	Human	15	11	4
	Mouse	15	8	7
GalNAc transferases	Human	12	3	9
	Mouse	8	1	7
Glc transferases	Human	3	3	–
	Mouse	1	–	1
GlcNAc transferases	Human	13	8	5
	Mouse	5	3	2
Man transferases	Human	10	6	4
	Mouse	3	1	2

further, and this should be considered when designing probes. From a survey of the CAZY and GenBank databases, it appears that for human, out of the 80 sequences required for probe design, 30 lack or have incomplete 3′-UTRs. In addition to the lack of 3′-UTRs for 27 out of 57 mouse sequences, a considerable number of murine orthologues have not yet been cloned or are not publicly available. A general effort is therefore needed in order to be able to construct an exhaustive glycogenes microarray covering both human and mouse species.

References

1. Varki, A. (1993) Glycobiology **3**, 97–130
2. Chee, M., Yang, R., Hubbell, E., Berno, A., Huang, X.C., Stern, D., Winkler, J., Lockhart, D.J., Morris, M.S. and Fodor, S.P. (1996) Science **274**, 610–614
3. Schena, M., Shalon, D., Davis, R.W. and Brown, P.O. (1995) Science **270**, 467–470
4. Chambers, J., Angulo, A., Amaratunga, D., Guo, H., Jiang, Y., Wan, J.S., Bittner, A., Frueh, K., Jackson, M.R., Peterson, P.A. et al. (1999) J. Virol. **73**, 5757–5766
5. Stingley, S.W., Ramirez, J.J., Aguilar, S.A., Simmen, K., Sandri-Goldin, R.M., Ghazal, P. and Wagner, E.K. (2000) J. Virol. **74**, 9916–9927
6. Schena, M. (ed.) (2000) Microarray Biochip Technology, Eaton Publishing, Natick, MA
7. Ekins, R.P. and Chu, F. (1994) Trends Biotechnol. **12**, 89–94
8. Hughes, T.R., Mao, M., Jones, A.R., Burchard, J., Marton, M.J., Shannon, K.W., Lefkowitz, S.M., Ziman, M., Schelter, J.M., Meyer, M.R. et al. (2001) Nat. Biotechnol. **19**, 342–347
9. Cameron, H.S., Szczepaniak, D. and Weston, B.W. (1995) J. Biol. Chem. **270**, 20112–20122
10. Van Gelder, R.N., von Zastrow, M.E., Yool, A., Dement, W.C., Barchas, J.D. and Eberwine, J.H. (1990) Proc. Natl. Acad. Sci. U.S.A. **87**, 1663–1667
11. Sekine, M., Nara, K. and Suzuki, A. (1997) J. Biol. Chem. **272**, 27246–27252
12. Igdoura, S.A., Gafuik, C., Mertineit, C., Saberi, F., Pshezhetsky, A.V., Potier, M., Trasler, J.M. and Gravel, R.A. (1998) Hum. Mol. Genet. **7**, 115–121
13. Kurosawa, N., Inoue, M., Yoshida, Y. and Tsuji, S. (1996) J. Biol. Chem. **271**, 15109–15116
14. Paulson, J.C., Weinstein, J. and Schauer, A. (1989) J. Biol. Chem. **264**, 10931–10934
15. Lee, Y.C., Kurosawa, N., Hamamoto, T., Nakaoka, T. and Tsuji, S. (1993) Eur. J. Biochem. **216**, 377–385
16. Coutinho, P.M., and Henrissat, B. (1999) in Recent Advances in Carbohydrate Bioengineering (Gilbert, H.J., Davies, G., Henrissat, B. and Svensson, B., eds.), pp. 3–12, The Royal Society of Chemistry, Cambridge

Biochem. Soc. Symp. **69**, 143–160
(Printed in Great Britain)

12

Neoglycoconjugates: trade and art

Nicolai V. Bovin[1]

Shemyakin-Ovchinnikov Institute of Bio-organic Chemistry, Russian Academy of Sciences, ul. Miklukho-Maklaya 16/10, Moscow GSP-7, 117997, Russia

Abstract

This chapter deals with the tendencies in design of multivalent neoglycoconjugates for glycobiology research and high-throughput profiling technologies, including cellular phenotyping. Soluble polyacrylamide (PAA) conjugates are remarkable owing to a variety of possibilities for synthesis and application. PAA is soluble and stable, and the molecule is flexible, PAA-tethered ligands are capable of adjusting to a receptor during multiple-point interactions and PAA does not bind to the cell surface. Synthesis provides unlimited diversity of the probe types (biotin, fluorescein, allyl, digoxygenin, ^{3}H, radiolabelled I), glyco-particles, glyco-surfaces, multiarrays, immunogens etc. Several examples illustrate the most advanced applications. (i) Dynamic systems: the selectin ligands immobilized on the surface as sugar–PAA conjugates made the study of the kinetics of rolling in the model system possible. Carbohydrate ligands that are covalently attached to the chip as sugar–PAA conjugates are of use with the surface plasmon resonance method. (ii) Pseudoglycoprotein: some questions arise regarding the biologically active glycoproteins, e.g.: is a carbohydrate or peptide fragment responsible for the activity? We have proposed the approach that promotes to answer this and other questions. The pool of oligosaccharides that were spitted off a glycoprotein is attached to PAA resulting in a pseudoglycoprotein. (iii) Virtual (dynamic) glycotope: receptor–ligand recognition, such as that of P-selectin with its receptor P-selectin glycoprotein ligand 1, frequently involves molecular interactions at two distinct sites. Using P-selectin as a model, we developed an approach to discover novel ligands. PAA was synthesized with multiple ligands ; a marked synergistic inhibitory effect was observed.

[1]E-mail: bovin@carb.siobc.ras.ru

Introduction

Carbohydrates involved in cell–cell recognition have been implicated in common human diseases. Therefore, studies in glycobiology provide a potential basis for the design of novel therapeutics. The development of carbohydrate-based therapeutic strategies is just beginning [1], but it is a slow process. Attempts to design such strategies are complicated by the peculiarities of glycoconjugates and by carbohydrate-mediated interactions. Problems that must be dealt with include glycan heterogeneity, weak affinities and multivalency of glycan–receptor interactions [2,3], and the fact that interactions between moving cells can differ from the static interactions. The search for target receptors has traditionally been based on studies of function, but recently many targets are identified as genes. For example, new members of the galectin and siglec families have been discovered via the Human Genome Project; genome analysis has revealed unknown lectin-like sequences. Proteins containing such sequences need to be characterized, with the goal of defining their carbohydrate-binding or carbohydrate-transforming abilities. This type of characterization demands a reproducible, robust and high-throughput approach. To achieve this goal and in order to address other issues in functional glycomics (see http://glycomics.scripps.edu), neoglycoconjugates have been developed as research tools. This chapter deals with current trends in the application of multivalent neoglycoconjugates to high-throughput profiling technologies (including cellular phenotyping) and in detailed studies on known carbohydrate-binding proteins.

Polyacrylamide (PAA) neoglycoconjugates

Before discussing practical applications, it is useful to review the properties of synthetic PAA-based neoglycoconjugates.

Need for multivalent probes

As a rule, the interaction of carbohydrate receptors or counter receptors does not correspond to the classic lock-and-key model. The low affinity constants for monovalent protein–glycan interactions must be compensated by multipoint interaction. Multivalent probes are well suited to the study of multipoint recognition. Ideally, neoglycoconjugates need to be synthesized not only as multivalent structures, but also of any molecular mass, size and flexibility, and should contain an unlimited repertoire of labels. Soluble PAA conjugates [4,5] are particularly useful because of the variety methods by which they can be synthesized and their analytical applications.

Advantages of PAA as a matrix for neoglyconjugate probes

In order to formulate the requirements for an ideal neoglycoconjugate, it is important to define what properties this tool should possess and what should be avoided in its design. The most rigorous requirements are for probes used for cell profiling, due to a high risk of non-specific interactions. Probes to be used in such studies need to be as hydrophilic as possible, but should be uncharged or slightly negatively charged to avoid interaction with cellular proteins and membrane components. The polymer matrix, spacer and label must all meet these conditions. The matrix providing multivalency should be flexible, so that several ligands are easily adjusted in space to interact with binding sites of proteins that consist of several subunits.

Based on these criteria, probes such as BSA derivatized with monosaccharides through phenyl spacers and labelled with biotin have characteristics that would make them unsuitable for the study of cells. The matrix is rigid and charged, the spacer and label are hydrophobic, and the carbohydrate ligand is small compared with the spacer group. Unfortunately, in spite of these shortcomings, such probes are often used. In contrast, PAA-based probes can be designed with much more desirable characteristics. Importantly, the polymer itself is soluble in water, has chemical and thermal stability and is resistant to enzymic degradation. The PAA matrix is a random coil, so the molecule is flexible and attached carbohydrate groups are free to change their positions relative to each other, before being fixed by a receptor. Thus PAA-tethered ligands, unlike those attached to rigid BSA molecules, are capable of adjusting to the receptor during multiple-point interactions. Moreover, owing to its lack of charge and the absence of hydrophobic groups, PAA itself does not bind to cell surfaces, as demonstrated by the interaction of tritium-labelled probes with Chinese hamster ovary cells that express E-selectin [6]. A small level of background was observed when either intact or permeabilized cells were assayed, which indicates that cell surface and intracellular components do not interact with PAA. Hydrophobic spacer arms, either aromatic or long aliphatic groups, can also cause non-specific probe–cell interactions. An advantage of the PAA matrix is that, owing to its high flexibility, it serves as a spacer. Therefore, a short $(CH_2)_2$ or $(CH_2)_3$ spacer is enough to allow binding of carbohydrate ligands to a protein.

Labels

Because PAA essentially causes no background, labelling groups attached to sugar–PAA probes can be the major cause of background binding. By using ELISAs to measure the binding to E-selectin expressed on the surface of Chinese hamster ovary cells, experiments were performed to find a label for sugar–PAA probes that provides an optimal signal/noise ratio [6]. Biotin and digoxygenin cause substantial increases in background binding compared with a tritium label. In contrast, probes containing a fluorescein label display a satisfactory background level (Figure 1). Binding of sialyl Lewis^x–PAA–fluorescein to the cells is dose-dependent and can be inhibited by free sialyl Lewis^x. In cell ELISAs, flow cytometry, fluorescence and confocal microscopy experiments [6–10], only tritium-labelled oligosaccharide–PAA and oligosaccharide–PAA–fluorescein probes really meet the strict requirements for cytochemical studies.

Synthesis

In addition to the probe structure, the synthetic pathway is also important. A novel approach to synthesis of PAA-based neoglycoconjugates

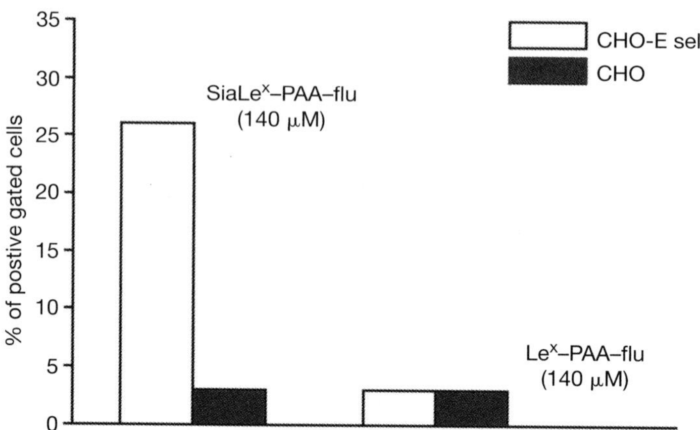

Figure 1 Flow cytometry analysis of sialyl Lewis^x–PAA–fluorescein binding to E-selectin-transfected and mock-transfected Chinese hamster ovary cells. The control used was the Lewis^x–PAA–fluorescein probe. CHO, Chinese hamster ovary cells; CHO-E, sel E-selectin-transfected CHO cells; SiaLe^x–PAA–flu, sialyl Lewis^x–PAA–fluorescein.

based on the coupling of the activated polymer with sugar amines (Figure 2) provides unlimited diversity of probe types and is also very practical [4,5]. The advantages of the approach, in comparison with co-polymerization of two or more monomers [11–13], are simplicity, consistency, reproducibility and unlimited potential for scaling up. Synthesis has proven to be very simple; it is accomplished in a moderate temperature range of 20–40°C, with routine reagents such as dimethylsulphoxide, dimethylformamide, triethylamine and 2-ethanolamine, and the sugar–PAA product is easily isolated.

Owing to the simplicity of the protocol and the quantitative yield, the product composition is highly reproducible on scales ranging from 0.1 mg to 10 g. The initial activated polymer is stable for a long time and is available in any amount. Thus, by using a single batch of the polymer, it is possible to obtain sugar–PAA samples of identical composition and molecular mass over a period of many years. The same is true for synthesis of more complex conjugates containing multiple types of sugars as well as labels. Coupling of different sugar amines to samples from a single batch of the polymer gives rise to a series of probes with absolutely identical molar amounts of sugar and label. Probes with exactly comparable sugar content are essential in studies that use sets of conjugates to analyse the specificity of carbohydrate-binding proteins. In addition, conjugates containing, for instance, 2, 5, 10, 15, 20 or 25 mol% sugar facilitate the selection of the reagent with the optimal sugar content. It is important to note that optimization does not always lead to the anticipated results, in the sense that greater sugar loading does not always lead to better binding to lectins [8].

Another major advantage of the PAA approach is that a great variety of conjugates can be synthesized using the same chemistry (Figure 3). Using the same method, different labels and effectors can be attached to the PAA

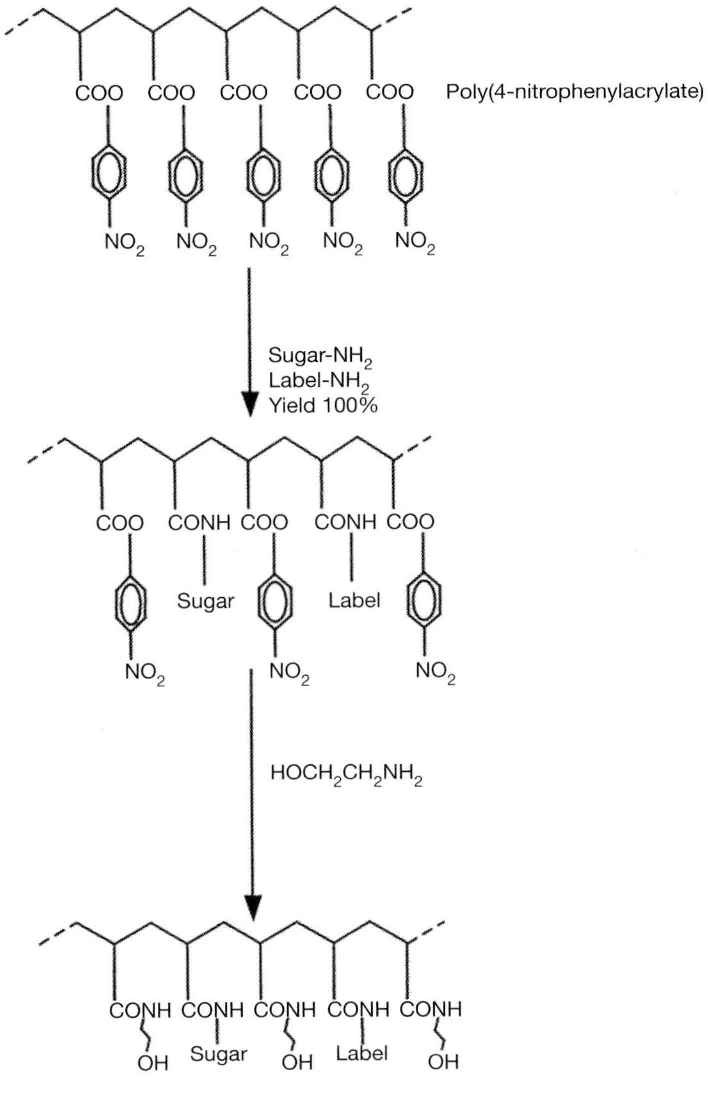

Figure 2 Scheme of PAA-based glycoconjugate synthesis. First stage: amino-containing ligand (sugar-NH_2, label-NH_2) attachment to activated polymer. Second stage: carrier modification resulting in final conjugate.

matrix. Moreover, sugars are only one type of ligand that can be attached to PAA. The method can also be applied to other ligands such as glycopeptides and peptides. In most of the studies cited in this review, 3-aminopropyl spacer groups are used. A short aminopropyl or even an aminoethyl linker is sufficient for the interaction of an oligosaccharide with lectins, antibodies or glycosyltransferases. Glycosylamines (without a spacer arm) and their

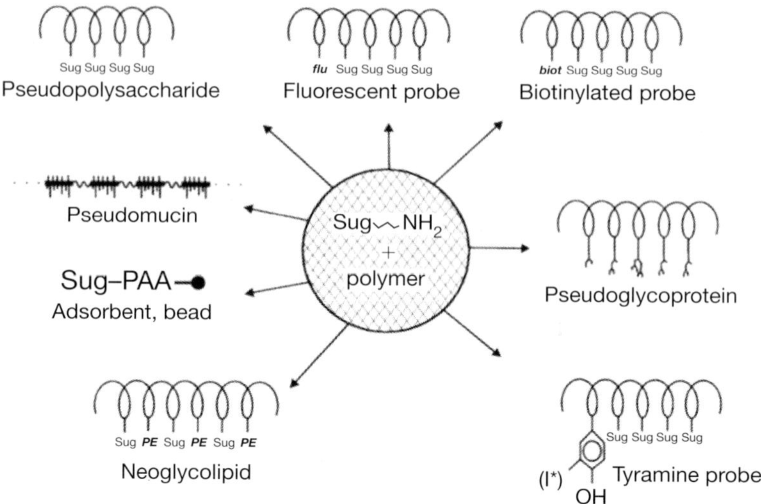

Figure 3 Diversity of PAA probes. The same procedure is used for preparation of probes bearing different labels: biotin (biot), fluorescein (flu), radioactive iodine, phosphatidylethanolamine (PE) or a particle. Pseudomucin consists of alternating glycosylated and unglycosylated motifs. A pseudoglycoprotein is obtained by attachment of an oligosaccharide pool released from a glycoprotein. Sug, sugar.

N-glycyl derivatives, sugar–$NHCOCH_2NH_2$, have provided good results [14].

The relative molecular masses of conjugates that have been prepared range from 30 to 2000 kDa. Normally, poly(4-nitrophenylacrylate) used as an activated precursor is obtained with a molecular mass of approx. 30 kDa, whereas that of an *N*-hydroxysuccinimide-activated polymer is approx. 2 MDa [15].

Application: trade

Sugar–PAA

Label-free sugar–PAA conjugates are used as inhibitors of various carbohydrate-mediated interactions. Compared with monovalent saccharides, the sugar–PAA conjugates are usually required at much lower concentrations. They can also serve as convenient coating reagents for immunological plates and nitrocellulose membranes in different solid-phase assays [4]. Sugar–PAA conjugates are applied not only for studies of antibodies and lectin, but also for glycosyltransferase [16] and glycosidase [17] assays. The radioactive labelling of these conjugates opens up additional possibilities. Synthesis of labelled conjugates only requires addition of labelled amino acid or another source of radioactivity into the reaction mixture, together with the sugar amine. Tritiated sugar–PAAs have been used in scintillation-proximity-assay-based high-throughput screening of selectin blockers [18], for the control of solid-phase coating with sugar–PAAs, and in cytochemistry (see above). Tyramine-containing probes have also been synthesized to allow introduction of radioactive iodine [19].

Sugar–PAA–biotin

The specificities of anti-carbohydrate antibodies can be investigated when sugar–PAA serves as a capture agent (see above). In the case of lectins, which often bind rapidly and non-specifically to a solid phase, a reversed version is preferable, in which lectin is applied to the solid phase, while sugar–PAA–biotin acts as a tracer. In this format, assay noise is reduced and the signal is amplified by subsequent interaction of biotin and streptavidin. Although a higher content of biotin amplifies the signal, it may provoke a non-specific interaction because of its hydrophobic nature. Particular care is required for cytochemical and histochemical research (see above). The optimal biotin content (5 mol%) has been determined empirically.

The opportunities for constructing different supermolecular complexes with sugar–PAA–biotin are unlimited. It is only necessary to have the corresponding streptavidin-containing 'components', which are often available commercially. There are several examples of how this approach can be used. Addition of the streptavidin–peroxidase conjugate to sugar–PAA–biotin instantly yields a cross-linked soluble supermultivalent conjugate, thus significantly increasing the sensitivity of assays using this conjugate [20]. Streptavidin-coated beads that are modified with sugar–PAA–biotin can be used for magnetic separation of tumour cells or selection of hybridomas [21]. If the particles are composed of fluorescent material, or if PAA is also labelled with fluorescein, cell-surface lectins can be revealed with high sensitivity. Label-free, carbohydrate-linked beads are appropriate reagents for latex agglutination, so they were used to develop a very simple field assay for leprosy [22]. Only a few minutes are needed to prepare a carbohydrate affinity adsorbent with any required carbohydrate capacity from sugar–PAA–biotin and streptavidin–Sepharose. Glycochips or glycoarrays in custom configurations can be created using any high-throughput platform, in which dots or wells are precoated with streptavidin [23]. Importantly, when sugar–PAA–biotin is used, the nature of the biotin spacer group is not critical, while in monomeric sugar–spacer–biotin both the length and nature of the spacer are crucial.

Sugar–PAA–fluorescein

Flow cytometry in combination with fluorescent probes, such as sugar–PAA–fluorescein, broaden opportunities for studying cell lectins, particularly when working with living cells, quantitative evaluation of binding and inhibition and simultaneous monitoring of several markers. A fluorescein label at a concentration of 1 mol% is sufficient for detection, while the non-specific interaction with the cells is negligible at such a low concentration (1 mol% corresponds to several fluorescein residues per molecule of probe). Detection of the Vero cell galectins by a series of sugar–PAA–fluorescein probes serves as an example (Figure 4). The staining of intact cells is weak; however, after membrane permeabilization, intense binding to intracellular galectins is observed.

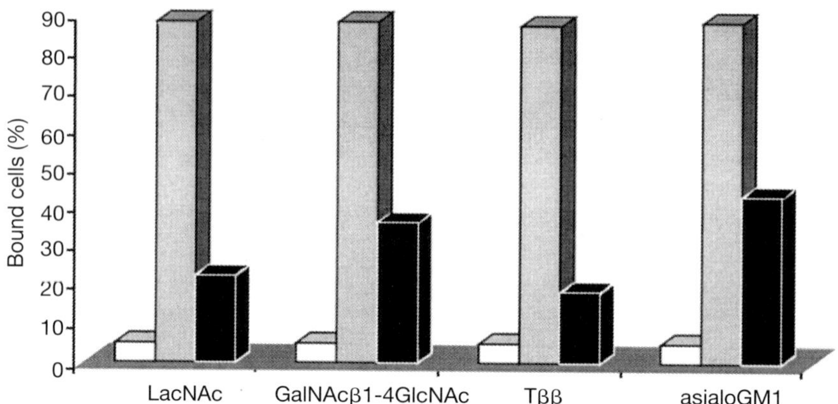

Figure 4 Flow cytometry analysis of galectin expression on Vero cells. The percentage of bound cells was calculated as $100-[(F_i/F_0)\times 100]$, where F_i is the fluorescence intensity of cells after incubation with sugar–PAA–fluorescein and F_0 is the fluorescence intensity of untreated cells. White bars, untreated cells; grey bars, cells after heat shock; black bars, cells treated with ionophore A23192. LacNAc, *N*-acetyl-lactosamine; Tββ, Galβ1–3GalNAcβ; asialoGM1, Galβ1–3GalNAcβ1–4 Galβ1–4Glc; GalNAc, *N*-acetylgalactosamine; GlcNAc, *N*-acetyl-lactosamine.

Sugar–PAA–fluorescein probes have been used to study recently described carbohydrate–carbohydrate recognition phenomena [24,25]. This biological interaction is calcium-dependent, extremely weak and displayed only in multivalent systems that involve cell–cell, cell–liposome, liposome–liposome or liposome–glycolipid interactions, and is seen only in conditions where the density of carbohydrate residues is elevated [26]. Therefore, multivalent probes bearing flexible and adjustable ligands could serve as tools for detecting and modelling carbohydrate–carbohydrate recognition. For example, mannose-containing PAA probes were used to demonstrate the interaction between yeast glucan and mannosylated glycoconjugates. The binding was sugar-, dose- and calcium-dependent, and could also be inhibited by small molecules that are related to both interacting carbohydrate partners.

Sugar–PAA–phosphatidylethanolamine (PE)

Neoglycolipids have been synthesized by the attachment of PE to PAA, in addition to the carbohydrate ligand. In some cases, a radioactive or fluorescent label has been attached to the polymer in addition to PE. This arrangement permits control of the introduction of lipophilic conjugates into the cell membrane using flow cytofluorometry. Lewis^y–PAA–PE was used as an immunogen to generate monoclonal antibodies that recognize a natural antigen on the cell surface [27]. The tailored design of a carbohydrate immunogen may lead to the generation of antibodies with different specificities such as those that recognize a cluster of closely spaced haptens.

Sugar–PAA–PE probes are suitable for incorporation into liposomes [9] and the cell membrane. To study the role of carbohydrate receptors during the natural killing process the probes were inserted into K562 cells. The incorporation was monitored using antibodies that were directed against the carbohydrate residue [28]. Expression of carbohydrate was observed for a few hours, which was enough time to study the promotion or inhibition of the killing of modified target cells by natural killer cells.

Application: from trade to art

In addition to the probes described above, new neoglycoconjugates have been developed recently. These new tools increase the possibilities for glycobiological studies.

Affinity PAGE

Carbohydrate ligands covalently cross-linked into a PAA gel can be used to detect and characterize carbohydrate-binding proteins. Conjugates with suitable stability and chemical reactivity for this purpose contain allyl groups, which are introduced into the polymer as $CH_2{=}CHCH_2NH_2$; the allyl group cross-links with the PAA gel matrix. This affinity gel is prepared as a thin layer within the stacking region of the PAA gel, and electrophoresis is performed under native, non-denaturing conditions (Figure 5). Carbohydrate-specific macromolecules, such as lectins, antibodies, aptamers or second saccharides that interact specifically with the immobilized sugar, are retarded during electrophoresis, which permits simultaneous separation of non-binding proteins according to size and charge [18].

Giant sugar–PAA

Conventional soluble PAA conjugates have an average molecular mass of approx. 30 kDa, which corresponds to a coi size of approx. 100 Å, based on data from electron microscopy. It is important to consider whether this size is sufficient for a multivalent mode of interaction with proteins. If the distance between the binding sites on the protein is less than the bead size, such interactions may occur, but if it is larger, either a weak interaction or no interaction will occur. In practice, multivalent interaction with widely spaced, sugar-binding sites can be achieved by creating significantly larger conjugates. Gel filtration data have shown that the maximum molecular mass that can be achieved is in the range 1–2 MDa, which corresponds to a coil size of approx. 500 Å, based on electron microscopy data. Molecules that are larger than this tend to form gels. The binding of these giant sugar–PAA conjugates to IgM is 10–100-fold stronger than that of common 30-kDa conjugates [20]. Similar enhancement of affinity was observed for 6′SLN–PAA–biotin (where 6′SLN is Neu5Acα2–6Galβ1–4GlcNAc) binding to the influenza virus. Both of these examples demonstrate that the distance between the binding sites in antibody or haemagglutinin surpasses the threshold value of 100 Å.

Dynamic systems

Selectin ligands that are immobilized on surfaces as sugar–PAA make the study the kinetics of rolling in a model system possible [29]. Carbohydrate

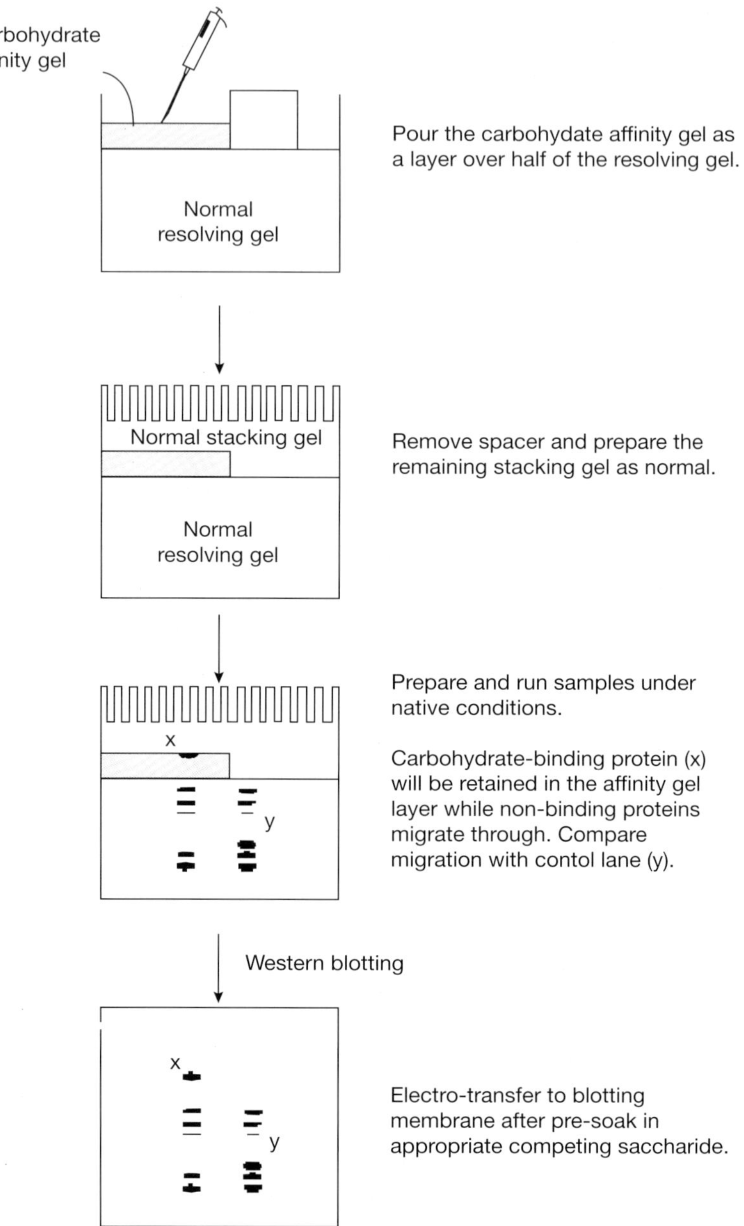

Figure 5 Affinity electrophoresis for lectin separation. Lectins are retained by the affinity layer, while non-binding proteins are separated as normal in the resolving gel. The affinity layer was prepared by co-polymerization of sugar–PAA–allyl with acrylamide.

ligands that are covalently attached directly to chips as sugar–PAA, or through streptavidin as sugar–PAA–biotin, are of use with the surface plasmon resonance method [30]. The kinetics of the protein–carbohydrate interaction for immobilized sugar–PAA and for sugars immobilized directly on to surfaces are significantly different. In the former case, the rate of protein binding is higher and elution occurs under milder conditions (D.K. Cooper, personal communication). Both of the features are advantageous for affinity chromatography.

Creation of glycolandscapes

Cell recognition frequently involves molecular interactions at two or more distinct sites [26,31]. In order to organize a number of glycans in a customized glycolandscape, a rigorous quantitative method for ligand-to-surface attachment would be of value. The present PAA technology is perfect for this purpose, as illustrated by the ability to modify surfaces with two different oligosaccharides. The goal of Khraltsova et al. [16] was to mimic the action of α-galactosyltransferase. For this purpose, sets of polymers simultaneously containing the disaccharide Galβ1–4GlcNAc (0–30 mol%) and the trisaccharide Galα1–3Galβ1–4GlcNAc (30–0 mol%) were synthesized; the total loading of disaccharide and trisaccharide was always 30 mol%. Using this series of polymers, the action of α-galactosyltransferase on Galβ1–4GlcNAc–PAA was mimicked. The polymers were coated on to 96-well plates and the α-Gal motif was detected by lectin binding. In parallel, Galβ1–4GlcNAc–PAA was applied on to a second plate and α-galactosyltransferase was added together with a galactosyl donor. The resulting enzymic modification was quantified using the same lectin. The resulting binding curves were identical (Figure 6), which indicates that the PAA-based approach to glycolandscape construction mimicked the transferase action.

The design of more complex PAA-based compositions seems possible. Figure 7 demonstrates a landscape combining a low-molecular-mass sugar–PAA as a glycolipid mimetic, a high molecular-mass sugar–PAA as a glycoprotein mimetic and pseudomucin which protrudes further than both of these [4]. This arrangement imitated not only carbohydrate heterogeneity, but also topography and accessibility of one or another landscape area to cells.

Pseudoglycoproteins

As a rule, a glycoprotein has several glycosylation sites and each of them is glycosylated heterogeneously. Some important issues arise regarding biologically active glycoproteins, such as whether a carbohydrate or peptide fragment is responsible for the activity, whether a glycan is involved directly in the peptide interaction or modulates it, and whether individual carbohydrate chains or the combination of several chains in a certain spatial arrangement are essential. We have proposed an approach that has the potential to address these issues [32]. In this approach, carbohydrate chains are completely released from the glycoprotein and the pool of oligosaccharides is attached quantitatively to PAA, which results in a pseudoglycoprotein, for example pseudo α_1-acid glycoprotein. Like natural α_1-acid glycoprotein, the resulting pseudo α_1-acid glycoprotein modulates interleukin and tumour necrosis factor production by blood cells [33]. Thus, a carbohydrate chain or several carbohydrate chains of α_1-acid glycoprotein act as a pharmacophore and the relative locations of the glycans do not play a role. This approach

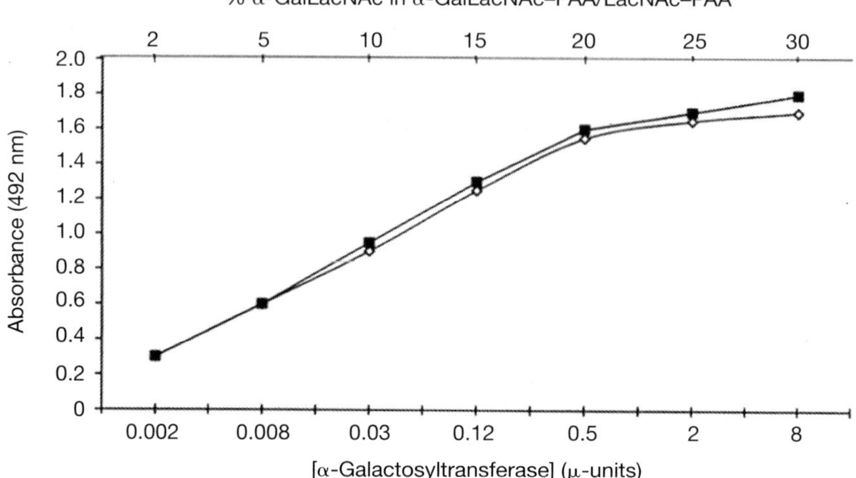

Figure 6 Glycolandscape. α-Galactosyltransferase assay. The curve ◇–◇ corresponds to the dependence of signal intensity (absorbance) on concentration of the enzyme added (lower horizontal axis). Plates coated with LacNAc–PAA was incubated with serially diluted α-galactosyltransferase in the presence of UDP–galactose. The curve ■–■ represents modelling the glycosylation landscape on the plastic surface by coating in various proportions of α-GalLacNAc–PAA/LacNAc–PAA (upper horizontal axis). LacNAc; N-acetyllactosamine; α-GalLacNAc, Galα1–3Galβ1–4GlcNAcβ.

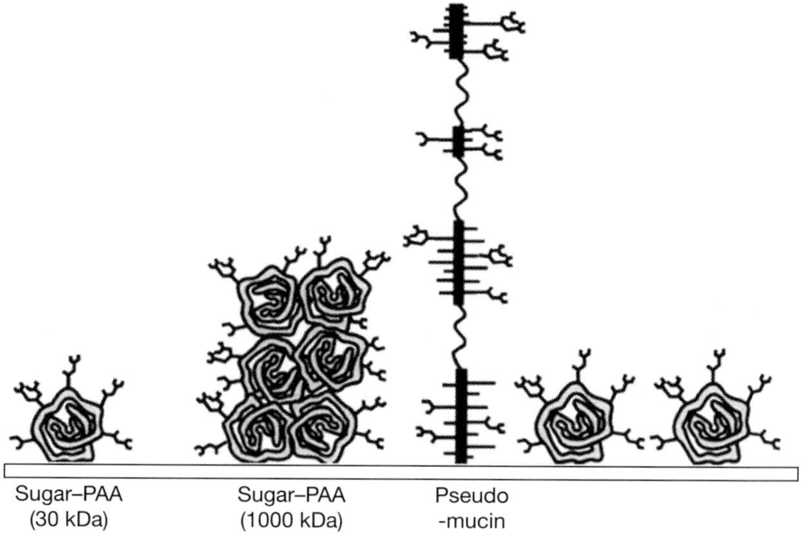

Figure 7 Glycolandscape mimicking the cell glycocalix. The real dimensions of different PAA glycoconjugates are represented.

might be employed to tackle more complex problems; for example, a chain can be selectively substracted from the pool, and its individual role can be traced.

Virtual combinatorial libraries

High-affinity receptor–ligand recognition frequently involves molecular interactions at two or more distinct sites. The interaction of P-selectin with its

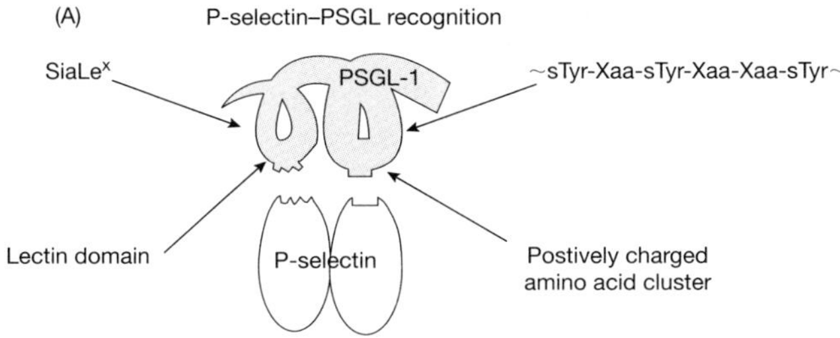

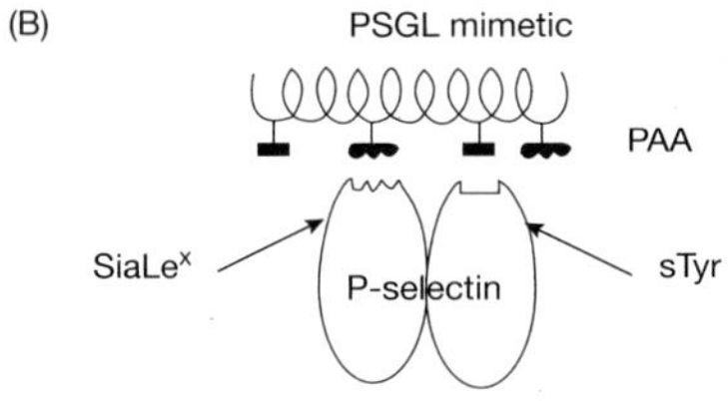

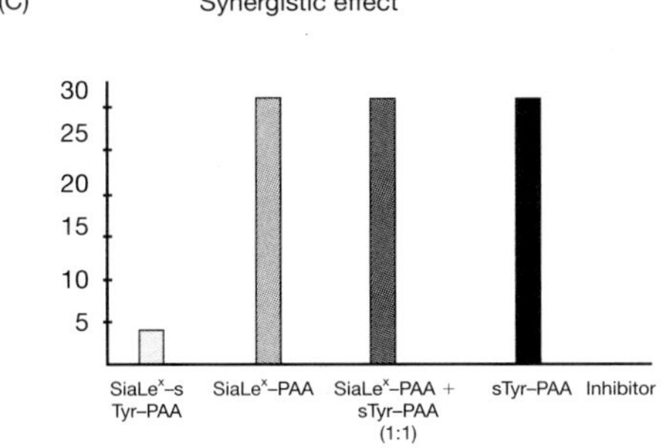

Figure 8 Design of P-selectin inhibitor. (**A**) Scheme of the two-site interaction of P-selectin with PSGL-1. (**B**) Design of a PSGL-1 mimetic by attachment of multiple copies of oligosaccharide and tyrosine sulphate to a flexible polymer PAA. (**C**) Synergistic effect of the sialyl Lewis^x/sulphotyrosine combination as an inhibitor of P-selectin. The inhibition of P-selectin binding with sulpho-Lewis^a–PAA–biotin was measured in a solid-phase assay. Each of the mono-ligand polymers, sialyl Lewis^x–PAA and sulphotyrosine–PAA, inhibits P-selectin with an IC_{50} of 30 μM. A mixture of the mono-ligand PAA conjugates has an additive effect only. The bi-ligand conjugate sialyl Lewis^x–PAA–sulphotyrosine has a higher inhibitory potency with an IC_{50} of 4 μM. sTyr, sulphated tyrosine. SiaLe^x, sialyl Lewis^x.

ligand P-selectin glycoprotein ligand 1 (PSGL-1) is as an example of this phenomenon. Primary interaction at site 1 is mediated through the lectin domain on P-selectin and sialyl Lewis^x-containing O-glycans on PSGL-1. A second, high-affinity interaction occurs between P-selectin and a cluster of sulphated tyrosine residues in the N-terminus of PSGL-1 (Figure 8A). Antagonists acting at either site may have potential anti-inflammatory effects. Using P-selectin as a model system, we developed a biomimetic approach to discover novel ligands [34]. A polymer was synthesized with multiple ligands on the backbone. Two-site P-selectin–ligand interactions were first studied by covalently incorporating sialyl Lewis^x and tyrosine sulphate on to the flexible polymer. In competition assays, a marked synergistic inhibitory effect was observed when a single polymer presented both ligands compared with polymers containing either ligand alone (Figure 8C). In a second approach, the sialyl Lewis^x ligand was reduced in complexity so that either Lewis^x or Lewis^a was attached to the polymer along with additional groups designed to mimic sialic acid. The combinations were better antagonists of P-selectin than sialyl Lewis^x itself [32].

Now that the basic principle of virtual combinatorial libraries as a strategy for ligand assembly has been established, the concept can be extended to broaden the types of functional group incorporated on to the polymer. The combinatorial approach using a multimeric matrix may be applied to discover ligand mimetics that are able to interact with a receptor or an enzyme. Many receptor–ligand interactions are complex, especially if the ligand is a macromolecule such as a protein or a complex glycan. The discovery of effective receptor antagonists has been hampered by the intricacy of these interactions. Standard black-box screening procedures for antagonists frequently target a single site on the receptor and ignore the potential for targeting multiple sites. Furthermore, the identity of the ligand interacting at a second site may be unknown.

Directed antigeneicity

The PAA matrix, unlike proteins such as serum albumin and keyhole limpet haemocyanin, is non-antigenic and non-immunogenic. Thus, one can design carbohydrate antigens without any risk of an immune response against peptide instead of carbohydrate epitopes. By using the PAA approach, we have attempted to develop a better understanding of the molecular basis of the blood group antigen AB. The AB specificity in blood group serology means that antibodies recognize the GalNAcα1–3(Fucα1–2)Gal (A) and Galα1–3(Fucα1–2)Gal (B) antigens equally, but do not interact with the shared epitope Fucα1–2Gal (H). We speculated [35] that the anti-AB antibodies would recognize the determinants A and B from the side opposite the C2′-NHAc group of trisaccharide A and the C2′-OH group of trisaccharide B (Figure 9); this idea was confirmed experimentally. The de-acetylated derivative of trisaccharide A was attached to PAA by the C2′ group rather than through the spacer at C1 as is normally the case, so that randomly coiled PAA would shield the natural antigenic site and present the epitope on the opposite side. The synthesized conjugate bound to monoclonal anti-AB antibodies, but lost its ability to bind anti-A and anti-B antibodies, for which the C2′-NHAc or C2′-OH group in the trisaccharide should essential.

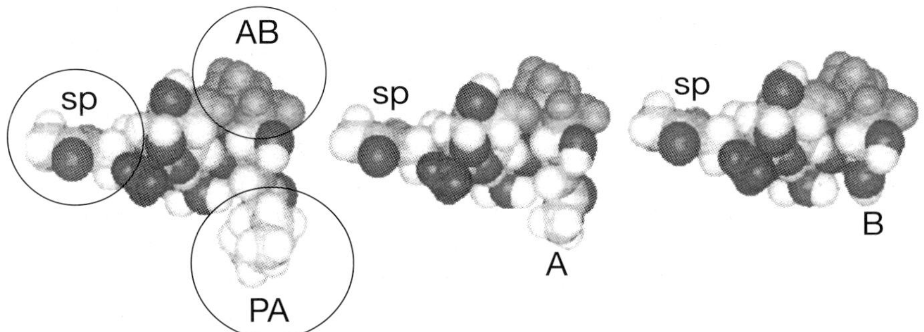

Figure 9 Molecular model of the AB trisaccharide coupled to PAA, and the glycotope responsible for the AB specificity in comparison with A and B specificities. The key groups of the AB epitope are on the opposite side to PAA. sp, spacer group.

Inspection of molecular models (Figure 9) permits us to speculate that anti-AB antibodies bind a cluster of closely spaced hydroxyl groups formed by the 2-OH of the fucose residue and the 4-OH and 6-OH of the α-galactose residue.

Glycolibraries

In conclusion, it is important to consider the most essential entities in glycoprobes, which are the carbohydrate ligands. For the purposes of glycomics, these must certainly be oligosaccharides rather than mimetics because there is no point in investigating an unknown object with the help of another, which is also unknown. Even a spacer group can significantly distort the probe–protein binding. For instance, an anti-sialyl Lewis^x antibody capable of recognizing the antigen on the cell surface interacts with Neu5Ac attached to an α-benzyl spacer much more strongly than it interacts with sialyl Lewis^x [36].

What repertoire of carbohydrate ligands should be included in glycome glycoarrays? When searching for and characterizing glycosyltransferases, sulphotransferases or O-acetyltransferases, the number of core chains that are potential acceptors is not very high. The problem is more substantial in the case of glycoarrays for mammalian lectin profiling; for example, the number of the known oligosaccharides that are potential ligands for siglecs approaches 100, while the number for galectins it is substantially higher. For collectins and ficolins, proteins of the innate immune system that recognize bacterial polysaccharides, the search for actual target molecules may require that screening be performed on a very large set of polysaccharides or fragments that number in the thousands.

Are there ways to create such comprehensive saccharide libraries? At first glance, rapid approaches such as solid-phase parallel and one-pot syntheses (reviewed in [37]) are rather attractive. However, attempts to develop oligosaccharide solid-phase synthesis always face the problem of separating extremely complex mixtures that result from low yields and incomplete stereoselectivity at each glycosylation stage. It is likely that only enzymic solid-phase synthesis stands a good chance of generating individual (pure) oligosaccharides. The same problem arises in cases of deliberate generation of complex mixtures, when a saccharide having several unprotected hydroxyl groups is glycosylated. Another strategy, chemical one-pot programmable oligosaccharide synthesis [37], requires that a diversity of synthons be used, each of which contains a thioglycoside group and one unprotected hydroxyl group. To synthesize a tetrasaccharide, for instance, four starting substances are mixed and thioglycoside groups are selectively activated one-by-one. An optimal choice of donor (–SR) and acceptor (–OH) group reactivity minimizes incorrect glycosylation at non-targeted hydroxyl groups. Like solid-phase synthesis, this approach has a disadvantage, which is the need to isolate a desired product from the very complex mixture of chromatographically related compounds. If the oligosaccharide is 90% pure, but an impurity has 10-fold higher activity than the major compound, the 10% contaminant can give a signal in bioassays that is comparable with the major compound. Such cases are well documented, as in the case of the epitope recognized by monoclonal antibody 101 [38].

In our conservative opinion, only two approaches exist for the creation of reliable libraries that meet the requirements for a glycomics array. The first involves a conventional step-by-step chemical or chemical/enzymic synthesis, in which each new compound is characterized carefully. The second involves use of oligosaccharides from such natural sources as milk and glycoproteins [39], but this approach must have associated impurity profiles and the complete characterization of structure. In this approach, two technical problems arise: separation of complex mixtures and lack of a universal spacering method. Nevertheless, it has been successfully realized in the well-known neoglycolipid technology [40].

Finally, the establishment of international research consortia (http://glycomics.scripps.edu) for generating comprehensive libraries is a positive advance in this area and will help to move towards the successful development of glycoarrays.

Work described in this chapter was supported by grants INTAS 97-32036, SCOPES 7SUPJ062207, and TSRI 1U54GM62116-01A1.

References

1. Hodgson, J. (1995) Biotechnology **13**, 38–39
2. Lee, R.T. and Lee, Y.C. (1994) in Neoglycoconjugates: Preparation and Application (Lee, Y.C. and Lee, R.T., eds), Academic Press, London
3. Hanson, J.E., Sauter, N.K., Skehel, J.J. and Wiley, D.C. (1992) Virology **189**, 525–533
4. Bovin, N.V. (1998) Glycoconj. J. **15**, 431–446
5. Bovin, N.V., Korchagina, E.Y., Zemlyanukhina, T.V., Byramova, N.E., Galanina, O.E., Zemlyakov, A.E., Ivanov, A.E., Zubov, V.P. and Mochalova, L.V. (1993) Glycoconj. J. **10**, 142–151
6. Galanina, O.E., Tuzikov, A.B., Rapoport, E.M., Le Pendu, J. and Bovin, N.V. (1998) Anal. Biochem. **265**, 282–289
7. Galanina, O., Hallouin, F., Goupille, C., Bovin, N. and Le Pendu, J. (1998) Int. J. Cancer **76**, 136–140

8. Gordeeva, E.A., Tuzikov, A.B., Galanina, O.E., Pochechueva, T.V. and Bovin, N.V. (2000) Anal. Biochem. **278**, 230–232
9. Vodovozova, E.I., Moiseeva, E.V., Grechko, G.K., Gayenko, G.P., Nifant'ev, N.E., Bovin, N.V. and Molotkovsky, J.G. (2000) Eur. J. Cancer **36**, 942–949
10. Galanina, O., Feofanov, A., Tuzikov, A.B., Rapoport, E., Crocker, P.R., Grichine, A., Egret-Charlier, M., Vigny, P., Le Pendu, J. and Bovin, N.V. (2001) Spectrochim. Acta Part A **57**, 2285–2296
11. Chernyak, A.Y., Antonov, K.V., Kochetkov, N.K. and Padyukov, L.N. (1985) Carbohydr. Res. **141**, 199–212
12. Kallin, E., Lonn, H., Norberg, T. and Elofsson, M. (1989) J. Carbohydr. Chem. **8**, 597–611
13. Bovin, N.V., Ivanova, I.A. and Khorlin, A.Y. (1985) Bioorgan. Khim. **11**, 662–670
14. Likhosherstov, L.M., Novikova, O.S., Derevitskaja, V.A. and Kochetkov, N.K. (1986) Carbohydr. Res. **146**, C1–C4
15. Lees, W.J., Spaltenstein, A., Kingery-Wood, J.E. and Whitesides, G.M. (1994) J. Med. Chem. **37**, 3419–3433
16. Khraltsova, L.S., Sablina, M.A., Melikhova, T.D., Joziasse, D.H., Kaltner, H., Gabius, H.-J. and Bovin, N.V. (2000) Anal. Biochem. **280**, 250–257
17. Rabina, J., Pikkarainen, M., Myasaka, M. and Renkonen, R. (1998) Anal. Biochem. **258**, 362–368
18. Rye, P.D. and Bovin, N.V. (1998) Biotechniques **25**, 146–151
19. Kojima, S., Andre, S., Korchagina, E.Y., Bovin, N.V. and Gabius, H.-J. (1997). Pharm. Res. **14**, 879–886
20. Weitz-Schmidt, G., Stokmaier, D., Scheel, G., Nifant'ev, N.E., Tuzikov, A.B. and Bovin, N.V. (1996) Anal. Biochem. **238**, 184–190
21. Rye, P.D. and Bovin, N.V. (1997) Glycobiology **7**, 179–182
22. Dyachina, M.N., Lukin, Y.V., Zubov, V.P. and Bovin, N.V (1992) Int. J. Leprosy **60**, 575–579
23. Pochechueva, T.V., Galanina, O.E., Nifant'ev, N.E., Kornilov, A., Tuzikov, A.B. and Bovin, N.V. (2002) In Leucocyte Typing VII (Mason, D., ed.), pp. 168–174, Oxford University Press, New York
24. Bovin, N.V. (1996) Biochemistry (Moscow) **61**, 968–983
25. Michalchik, E.V., Shiyan, S.D. and Bovin, N.V. (2000). Biochemistry (Moscow) **65**, 494–501
26. Hakomori, S. (1994) in Complex Carbohydrates in Drug Research, (Bock, K. and Clausen, H., eds), pp. 337–349, Munksgaard, Copenhagen
27. Vlasova, E.V., Byramova, N.E., Tuzikov, A.B., Zhigis, L.S., Rapoport, E.M., Khaidukov, S.V. and Bovin, N.V. (1994) Hybridoma **13**, 295–301
28. Kovalenko, E.I., Sablina, M.A., Khaidukov, S.V., Khirova, E.V. and Bovin, N.V. (1998). Rus. J. Bioorgan. Chem. **24**, 200–203
29. Brunk, D.K. and Hammer, D.A. (1997). Biophys. J. **72**, 2820–2833
30. Hirmo, S., Artursson, E., Puu, G., Wadstrum, T. and Nilsson, B. (1998). Anal. Biochem. **257**, 63–66
31. Crocker, P.R. and Feizi, T. (1996) Curr. Opin. Struct. Biol. **6**, 679–691
32. Pochechueva, T.V., Galanina, O.E., Bird, M.I., Nifant'ev, N.E. and Bovin, N.V. (2002) Chem. Biol. **9**, 757–762
33. Shiyan, S.D. and Bovin, N.V. (1997). Glycoconj. J. **14**, 631–638
34. Game, S.M., Rajapurohit, P.K., Clifford, M., Bird, M.I., Priest, R., Bovin, N.V., Nifant'ev, N.E., O'Beirne, G. and Cook, N.D. (1998) Anal. Biochem. **258**, 127–135
35. Pochechueva, T.V., Shipova, E.V., Pasynina, G.V., Korchagina, E.Yu. and Bovin, N.V. in 4th International Workshop on Monoclonal Antibodies Against Human Red Blood Cells and Related Antigens Paris, 2001, http://www.ints.fr/4thworkshop/bin/axodoc_download.php?&id=146&

36. Vlasova, E., Kovalenko, E., Galanina, O., Nifant'ev, N. and Bovin, N. (1997) in Leucocyte Typing VI: White Cell Differentiation Antigens. Proceedings of the Sixth International Workshop and Conference (Kishimoto, T., ed.), pp. 670–671, Garland Publishing, New York
37. Koeller, K.M. and Wong, C.-H. (2000) Glycobiology **10**, 1157–1169
38. Le Pendu, J., Fredman, P., Richter, N.D., Magnani, J.L., Willingham, M.C., Pastan, I., Oriol, R. and Ginsburg, V. (1985) Carbohydr. Res. **145**, 347–349
39. Tuzikov, A.B., Gambaryan, A.S., Juneja, L.R. and Bovin, N.V. (2000) J. Carbohydr. Chem. **19**, 1191–1200
40. Feizi, T., Stoll, M.S., Chai, W. and Lawson, A.M. (1994) Methods Enzymol. **230**, 484–519

Subject index

β-N-acetylglucosaminidase, 9, 11, 12
N-acetylglucosaminyl transferase
  Core 1, 39, 44
  Core 2, 37–40, 42, 43
asialoglycoprotein receptor, 6, 64

bactriophage T4 β-glucosyltransferase, 24–30
biotin, 148, 149
BTG (*see* bactriophage T4 β-glucosyltransferase)

CACH (*see* childhood atxia with central nervous system hypomyelation)
CAD–MS (*see* collision-activated dissociation–MS)
Caenorhabditis elegans, 2, 6–16, 62, 67–69, 74, 76, 86, 118–120, 122, 126–131
calnexin, 73
calreticulin, 73
carbohydrate-recognition domain, 60, 61,64, 66, 68, 70, 74, 76
CAZY database, 24, 27, 140, 141
CD33-related siglec
  domain structure, 84
  expression pattern, 90, 91
  mouse, 88, 89
  sialic acid-binding, 85, 87, 90
childhood atxia with central nervous system hypomyelation, 113, 114
coagulation factor V, 78
coagulation factor VIII, 78
collision-activated dissociation–MS, 123–127
COP-I protein, 74–76
COP-II protein, 74–77
CRD (*see* carbohydrate-recognition domain)
C-type lectin-like domain (CTLD), 60–69

N-deacetylase/N-sulphotransferase (NDST), 49, 51–55
α-DG (*see* α-dystroglycan)
Dictyostelium discoideum, 132
Drosophila melanogaster, 2, 13, 62, 67–69, 86, 132
dystroglycan, 4, 13, 17
α-dystroglycan, 109, 110, 112
dystrophin–glycoprotein complex, 4

electrospray ionization–ion trap–MS, 7, 11
electrospray ionization–quadrupole–time-of-flight–MS, 13–16
endoplasmic reticulum, 73–77
endoplasmic reticulum–Golgi intermediate compartment protein of 53 kDa, 73, 74, 76, 77
    function of, 74, 75
    -like (*see* ERGL)
    -related proteins, 78–80
ER (*see* endoplasmic reticulum)
ERGIC-53 (*see* endoplasmic reticulum–Golgi intermediate compartment protein of 53 kDa)
ERGL, 79–81
ESI–IT–MS (*see* electrospray ionization–ion trap–MS)
ESI–Q–TOF–MS (*see* electrospray ionization–quadrupole–time-of-flight–MS)
extrachromosomal tandem array, 7

fast atom bombardment–MS (FAB–MS), 106, 119–121, 126
fluorescein, 148–150
Fringe, 3, 4
fucose, 122
fucosyltransferase, 40–44, 130

α-galactosyltransferase, 153, 154
galectin, 69, 131
  apoptosis, 98
  chondrogenesis, 99, 100
  embryonic implantation, 97, 98
  expression pattern, 96
  immune system, 99
  mouse embryogenesis, 97
  olfactory system, 98
GeneChip®, 136

GlcNAcT I (*see* UDP-N-acetyl-
glucosamine:α3-D-mannoside
β1,2-N-acetylglucosaminyl-
transferase I)
GlyCAM-1 (*see* glycosylation-dependent
adhesion molecule 1)
glycoarray, 149, 157, 158
glycolibrary, 157, 158
glycosaminoglycan, 2, 3, 29
glycosylation, 1–7, 14, 16
  congenital disorders of, 2
  -dependent adhesion molecule,
  1, 37–41
  effect of mutation, 5–12
glycosyltransferase, 23, 33, 34, 136–139
  fold-recognition studies, 27–30
  structures of, 24–26
Golgi apparatus, 3, 9, 51, 73–76, 78, 95
GT (see glycosyltransferase)

Haemonchus contortus, 120
heparan sulphate, 47–55
  polymerase, 49–51
heparin, 47, 48, 49, 52, 54

immunoreceptor tyrosine-based
inhibitory motif, 84, 85, 88, 91, 92
immunoreceptor tyrosine-based switch
motif, 84, 85, 92
InterPro database, 62, 86
ITIM (*see* immunoreceptor
tyrosine-based inhibitory motif)
ITIM (*see* immunoreceptor
tyrosine-based switch motif)

laminin, 4
lectin
  C-type, 59, 68, 69, 131
  early secretory pathway, 72, 73
  I-type, 59, 68, 69, 83
  L-type, 68, 69
  M-type, 68, 70
  R-type, 68, 70
  S-type (*see* galectin)

liquid chromatography–
electrospray–MS, 6
liquid chromatography–tandem MS, 7
lysosomal storage disease, 2, 3

mannose-binding domain, 64
mannose 6-phosphate receptor, 69
α-mannosidase II null mice, 107
matrix-assisted laser-desorption
ionization–time-of-flight–MS, 13
MECA-79 antibody, 40
O-methylated fucose, 130

microarray
  cDNA target, 138
  DNA probe, 137
  glycosyltransferase probe, 137, 138
  signal quantification, 138, 139
MS analysis, 6, 7, 11–13, 15, 106, 107,
119–121, 123, 124, 126, 127–129
MS/MS (*see* tandem MS)
mucolipidosis, 3
muscle–eye–brain disease, 4
myd mouse, 106, 107, 110–112

natural killer cell, 88, 91, 92
NDST (*see* N-deacetylase/N-sulpho-
transferase)
neuraminidase, 140
NMR, 7, 128
Notch, 3, 4, 13, 17

OST (*see* O-sulphotransferase)

PAA (*see* polyacrylamide)
paucimannose, 7–12, 13
PC (*see* phosphorylcholine)
peptide N-glycosidase, 119–129
Pfam database, 62
phosphatidylethanolamine, 148, 150
phosphorylcholine, 125–128
PNGase (see peptide N-glycosidase)
polyacrylamide,
  glycoconjugate synthesis, 145–147
  probe label, 148–150
  properties, 144–145
post-translational modification, 1, 2, 4–7,
13–15, 17
presenilin, 77, 79
PROSITE database, 62
proteoglycan, 2, 5
proteome, 5
proteomics, 5–7, 13, 15, 16
P-selectin glycoprotein ligand 1, 42–44,
156
pseudoglycoprotein, 153
PSGL-1 (see P-selectin glycoprotein
ligand 1)

quadrupole–time-of-flight–MS
(Q-TOF–MS), 123–127

Rossmann fold, 29, 30

Saccharomyces cerevisiae, 132
selectin, 3
  E-, 34–36, 40–44
  P-, 34–36, 40–44, 156
  L-, 34–36, 38–44
sequon, 4, 13, 15, 16

SHP, 91, 92
sialyl Lewis^x, 39–42, 156
siglec (*see also* CD33-related siglec), 69, 83
signalling lymphocyte activation molecule, 92
SLAM (*see* signalling lymphocyte activation molecule)
SMART database, 62
SpsA, 24–30
Src homology 2-domain-containing protein tyrosine phosphatase (*see* SHP)
sulphotransferase, 37, 40, 41, 43
O-sulphotransferase, 41, 51–53, 55

tandem MS, 7, 15, 16, 123, 124, 126, 127, 129
transgenic mouse, 5, 14
two-dimensional isoelectric focusing-SDS/polyacrylamide gel electrophoresis, 13–15

UDP-N-acetylglucosamine: α3-D-mannoside β1,2-N-acetylglucosaminyltransferase I, 2, 5, 8–12

vesicular integral membrane protein 36 (VIP36), 73, 74, 78–80

Worm Proteome Database, 14